Dedicated to Galit and Dana

Published by Thomasson-Grant, Inc.
First published by Edizioni White Star. Title of the original
edition: *I Colori del Blu.*
World copyright © 1990 by Edizioni White Star,
Via Candido Sassone 22/24, 13100 Vercelli, Italy.

Printed in Singapore.

97 96 95 94 93 92 91   5 4 3 2 1

Library of Congress Cataloging-in-Publication Data
Rotman, Jeffrey L.
     (Colori del blu.   English)
     Colors of the deep / photographs by Jeffrey L. Rotman ;
  text by Joseph S. Levine
          p.   cm.
     Translation of : I colori del blu.
     ISBN 0-934738-87-4
     1. Marine biology--Pictorial works. 2. Underwater photography
  I. Levine, Joseph S., 1951-   II. Title.
  QH91.17.R6813   1991
  574.92'022'2--dc20   91-29
                                             CIP

Thomasson-Grant, Inc.
One Morton Drive, Suite 500
Charlottesville, Virginia 22901
(804) 977-1780

COLORS OF THE DEEP

PHOTOGRAPHS BY JEFFREY L. ROTMAN

TEXT BY JOSEPH S. LEVINE

THOMASSON-GRANT
CHARLOTTESVILLE, VIRGINIA

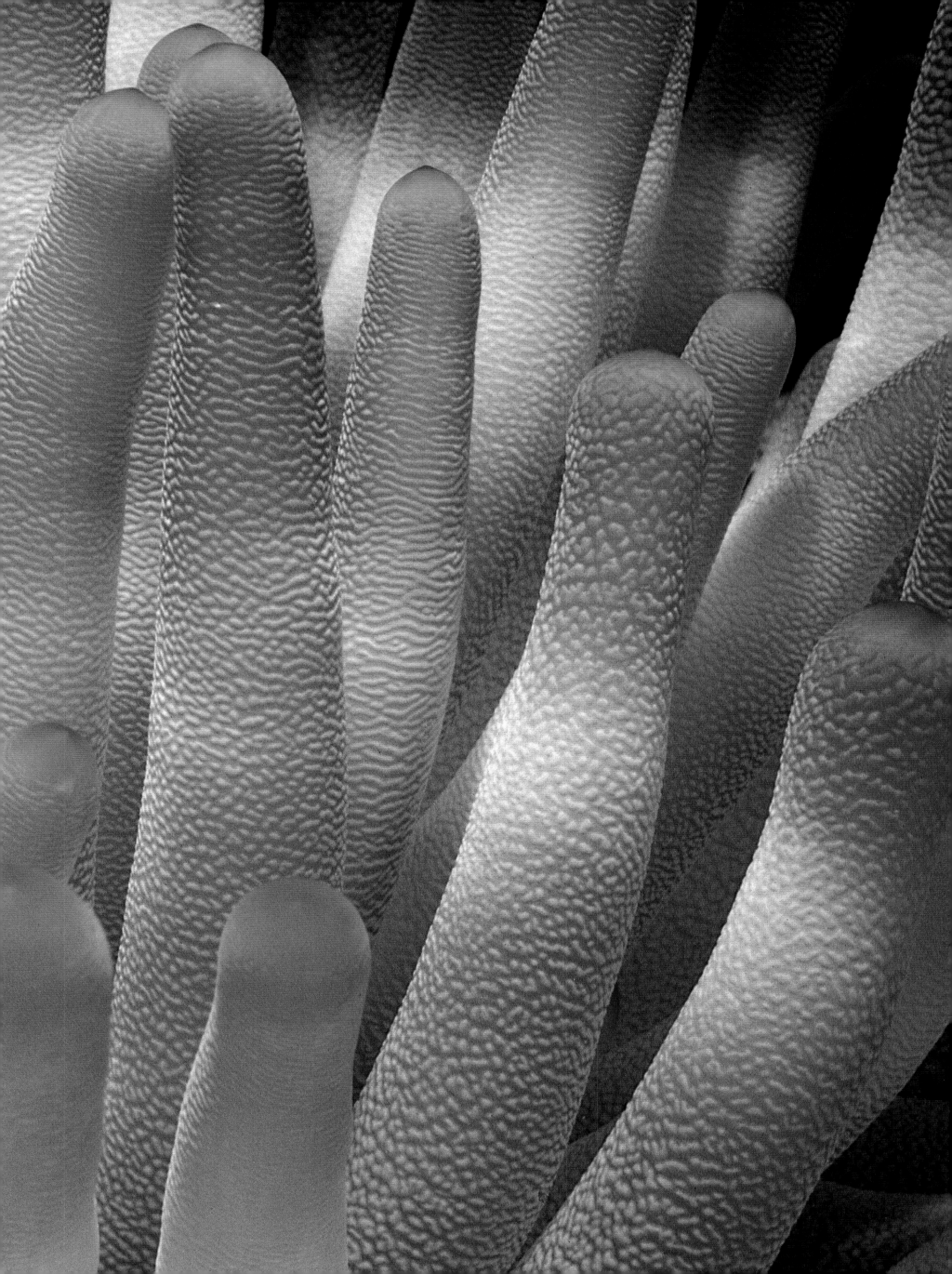

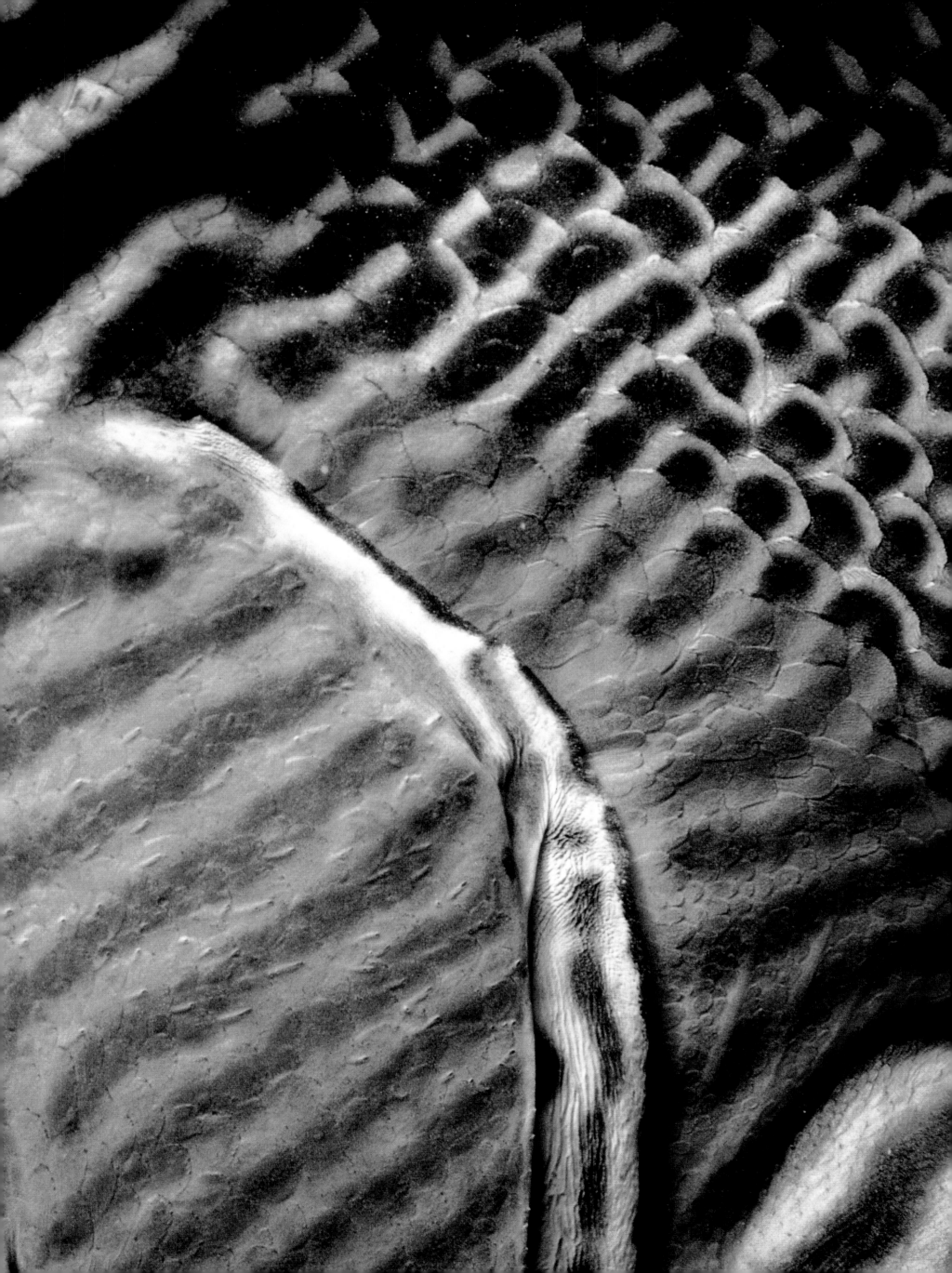

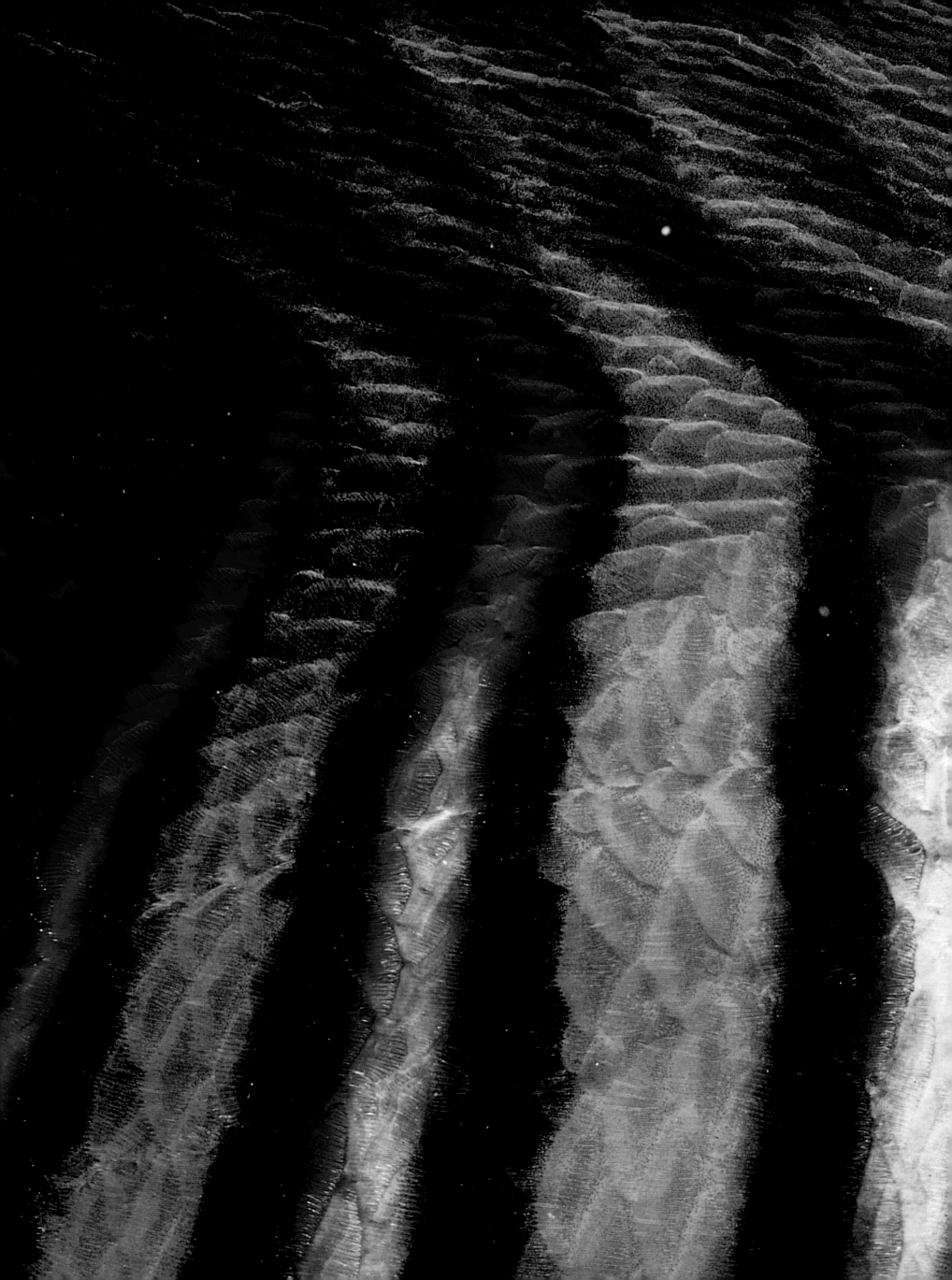

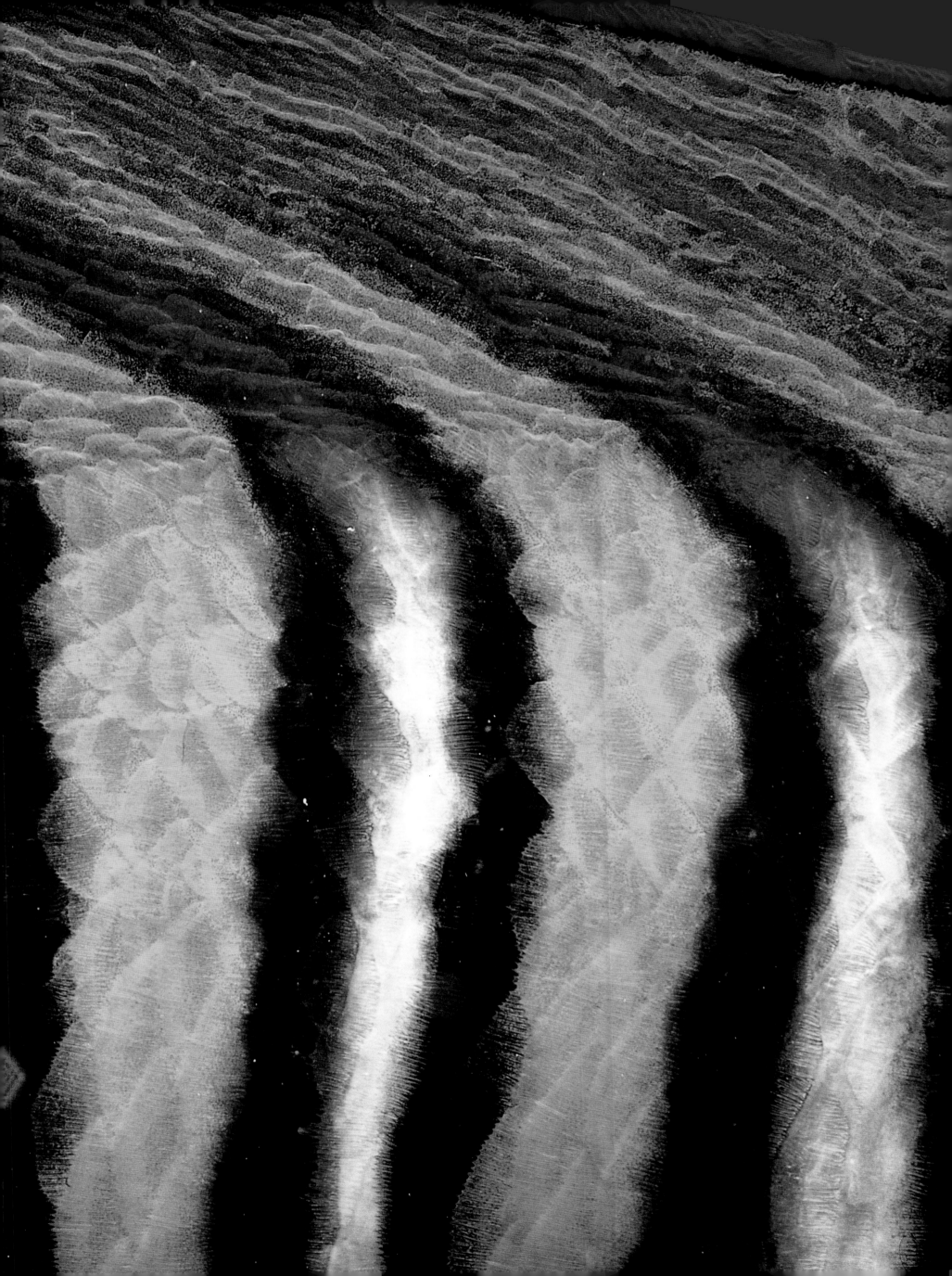

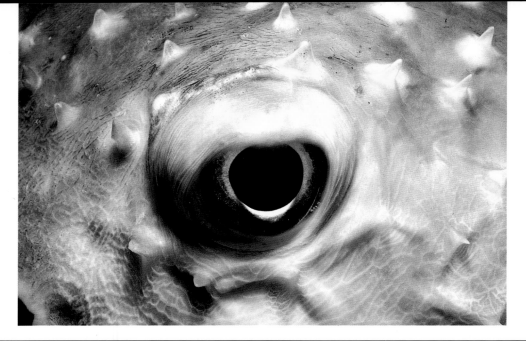

Left
Pufferfish, Red Sea.

Below
Jewel grouper, Red Sea.

Right
Parrotfish, Red Sea.

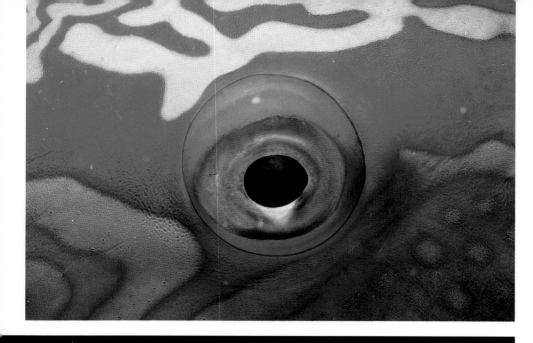

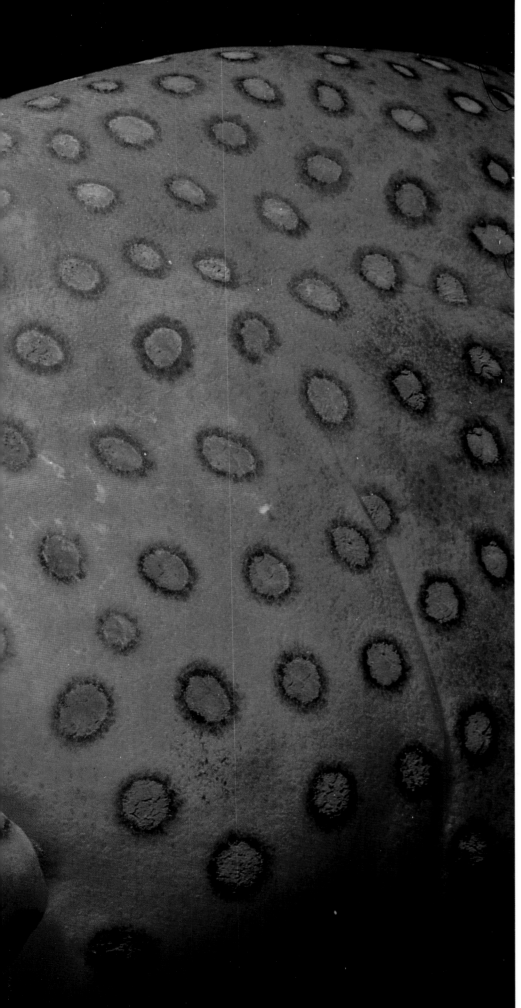

CONTENTS

Endsheets
Daisy coral, Red Sea.

Page 1
Parrotfish, Red Sea.

Pages 2-3
Dorsal fin of a parrotfish, Red Sea.

Pages 4-5
Sea anemone, Red Sea.

Pages 6-7
Jewel grouper, Red Sea.

Pages 8-9
Surgeonfish, Red Sea.

Pages 10-11
Royal angelfish, Red Sea.

When you first begin diving, it's natural to try to take in the wide expanse, the big picture. Later on divers often find themselves focusing on smaller subjects: individual fish, crabs, anemones, corals, and sponges that carpet this blue, fluid world. Fifteen years ago I started taking yet another step and began to look for that part of the whole which for me seemed to be the essence of the moment. Sometimes it was the eye of a sleeping fish or the mouth of an anemone. It could also be the design on the skin of the fish or the tentacles of the anemone falling gracefully across its body.

The ocean is not a still environment. It is alive and moving. With this movement images constantly change. Coral polyps pulsate, changing their shape every few seconds. The most gentle of currents causes the tentacles of anemones to swing and sway, presenting the "moment" when all your senses signal your shutter finger to record the image. At times that image was elusive. Perhaps my senses couldn't signal quickly enough, or maybe I was so lost in the image that I forgot to take the photograph altogether. But many of those images, however imperfect they may now seem, did transfer from my eye onto film.

Taking these images was an evolutionary process for me as a photographer and as a diver. I began with subjects that were attached to the substrate or, like many camouflaged fish, reluctant to move, thinking themselves invisible. Possibly my greatest development came when I began photographing at night in earnest. The cover of darkness allowed me to see but not be seen. Animals that stayed hidden during the daylight hours came out in force at night. Other animals, particularly the fish that moved too fast during the day and preferred to distance themselves from my lenses, became approachable by night. Some, peacefully asleep, allowed themselves to be handled gently. The black of night also increased my ability to concentrate on just what lay in front of my eyes. Schools of fish were certainly passing by, but the cloak of night made them much easier to forget.

The sea is never quick to reveal its secrets. The treasures I sought were small, frequently no larger than a thumbnail. Searching for my subjects in the North Atlantic or the Galápagos Islands, dark and often cold environments, required special patience and persistence. When you enter these steel-blue waters, they first seem poor in life. The fish quickly retreat, and a rich growth of algae and plants does well at hiding the invertebrate population. These are environments of camouflage that invite the curious and only begrudgingly reveal their secrets. But once the search begins, and you poke your nose into places it really doesn't belong, patience and perseverance are rewarded grandly.

Adding to these difficulties was the disappearance of colors. As sunlight penetrates deeper into the water, reds, oranges, and yellows are absorbed, attenuated, and scattered. Below 30 meters, where many of these photographs were taken, everything appears bluish-greenish-gray. Imagine my surprise at switching on my flashlight and seeing the electric orange of coral polyps at 75 meters. At these depths I already felt drunk from the nitrogen narcosis. The electric colors restored by my camera flash in these dark depths added to that wonderful feeling of fantasy.

In the ocean there are things that are beautiful, things that are bizarre, and a few that are both. Coming face-to-face with a hermit crab, focusing at one moment on the eyes that sit at the tips of stalks and then shifting attention to its hairy arms that almost seem cartoonish, is an exquisite experience. Pan-

ning across the dorsal fin of a sleeping parrotfish, trying to decide on the borders of the photographic frame, leads to seemingly timeless moments of blissful confusion. The way those colors would wash into one another abruptly and then begin again would fill my mind in much the same way a fine painting would. Science can explain why these colors are in their places, but the visual celebration of this palette of colors will never have for me a sufficient explanation, nor will it ever need one. It is because it is, and I delight in the hours of pleasure it gives me.

I am often asked what I consider to be the best dive location in the world. Invariably I name the Red Sea, the Galápagos Islands, the Caribbean, the Mediterranean, the North Atlantic, the Pacific Coast of North America, Micronesia, Hawaii, and Japan. Each environment is unique, and each offers something special. This book is my proof of that fact. The Red Sea hosts some of the most spectacularly colored coral reef fishes, but it would be hard-pressed to come up with a fish as arrestingly bizarre as the red-lipped batfish found in the waters encircling the Galápagos. While the soft corals of the Palau Islands in Micronesia are sure to knock your socks off, the ugly but beautiful face of the American goosefish seen in the coastal waters of New England's chilly North Atlantic can result in no less wonder. Whether swimming through the kelp forests of California or the soft coral jungles of the Izu Peninsula in Japan, I am constantly reminded how different, varied, and wonderful are the planet's salt-water environments.

If there is a purpose to these photographs, it is to share the excitement, fascination, and wonderful satisfaction I have experienced while taking them. Thousands of hours and images went into this project. The time spent never seemed to be an issue; the result always became more than a justification. While I was underwater, time often lost its usual measurement of hours and minutes.

Once, after spending an entire dive hunched over a particularly special anemone, I turned to see my safety diver visibly shaking from the cold. When we surfaced, he said that I had spent three hours and fifteen minutes with that one anemone. "Why?" he asked. I could only answer, "Because I was trying to see it." To my eye it offered many faces, a few smiles, and more than one wink. It never looked the same, and in its changes I saw the soul of that anemone, and perhaps the sea itself.

Jeffrey L. Rotman

Facing
Sand tiger shark, Florida Keys.

Pages 16-17
Coral polyps at rest, Red Sea.

Pages 18-19
Crab, Red Sea.

Pages 20-21
Clownfish and anemone, Red Sea.

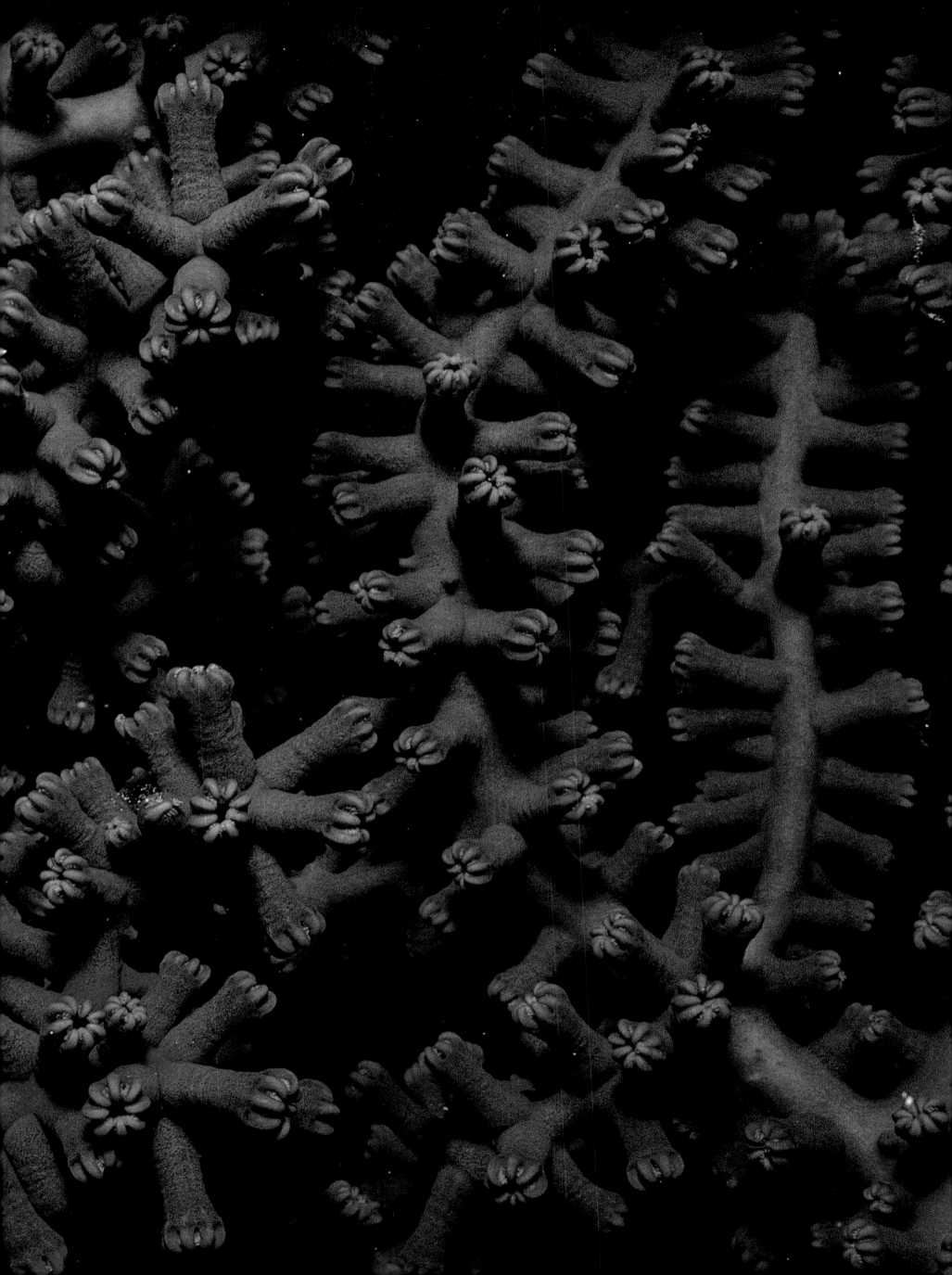

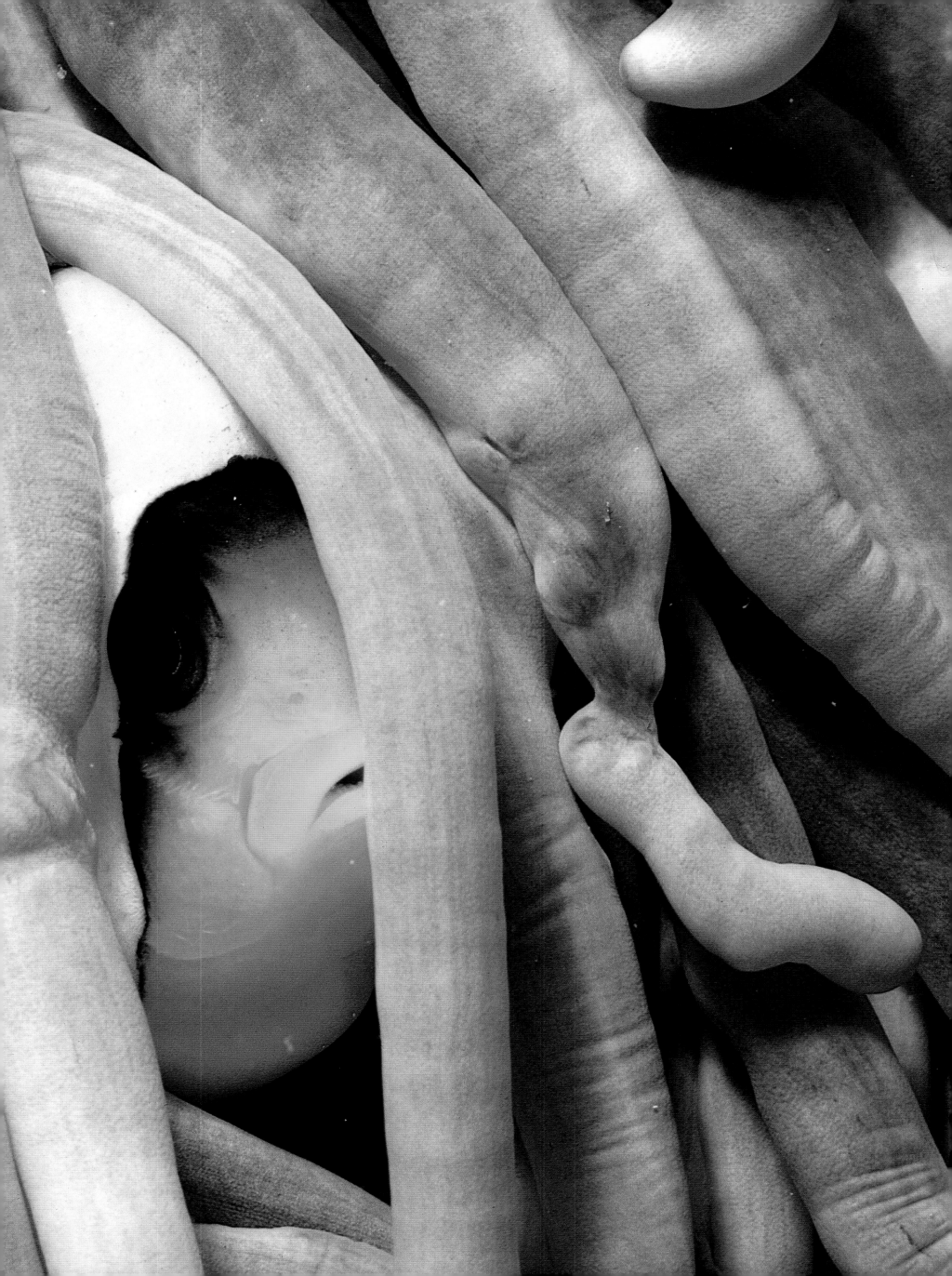

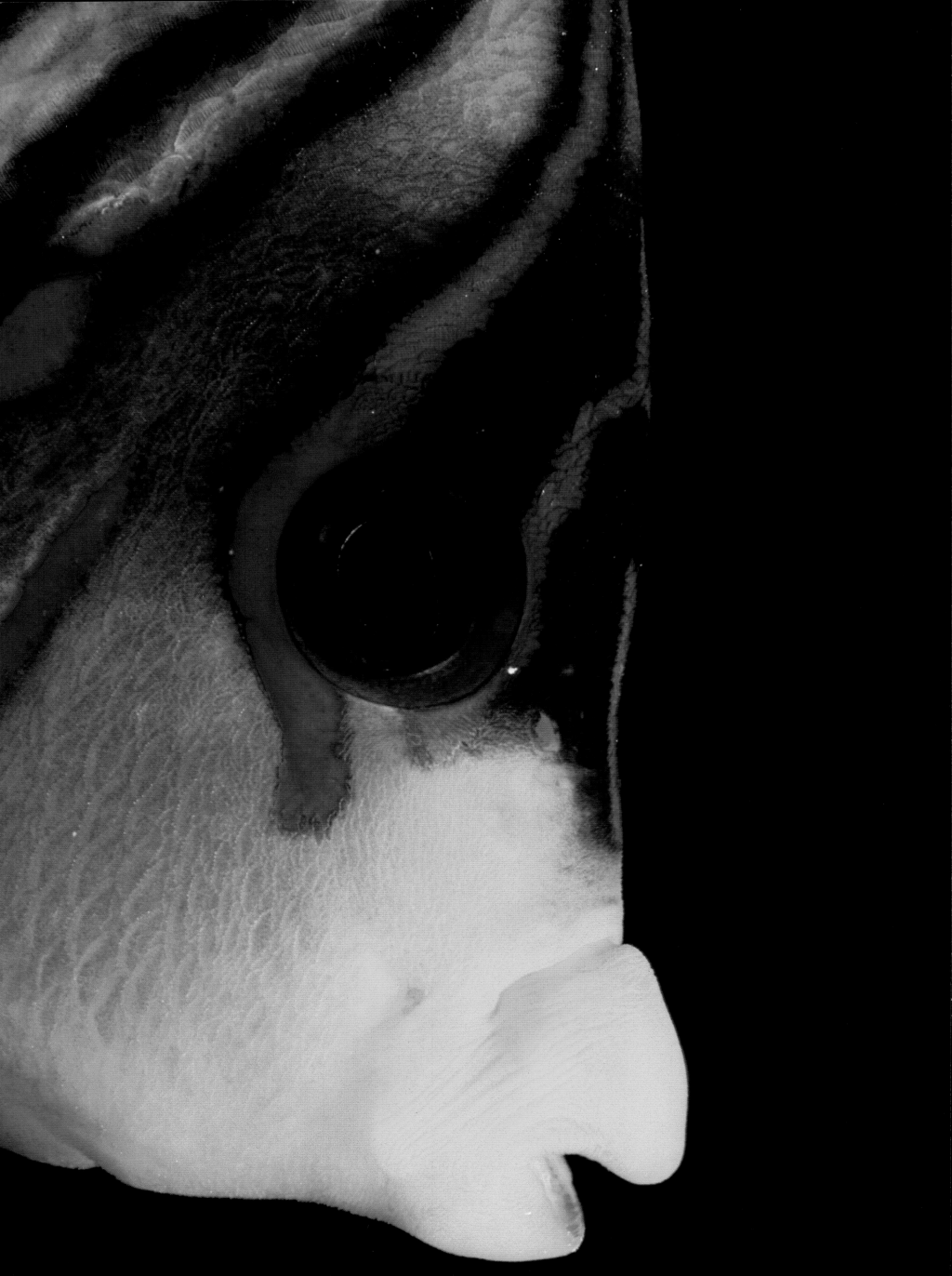

INTRODUCTION

Joseph S. Levine

Na'ama Bay, 7:30 A.M. The sun pulls free of the haze over Saudi Arabia to begin its low winter arc across the Gulf of Aqaba. As it rises it ignites images that challenge the accuracy of eye and mind.

Colors. Picture, in your mind's eye, a V-shaped valley walled with cliffs of granite and Nubian sandstone, airbrushed in subtle shades of orange and gray, and veiled in haze. Now imagine the deepest, purest blue of a mountain sky, distilled, intensified, liquefied, and poured into the depths of that valley. The juxtaposition evokes disbelief; the view cannot be real. The sea is too vivid and saturated, the mountains too ethereal. This must be a surrealistic painting or a photographer's mistake. But no blinking of the eyes or shaking of the head can restore a reasonable balance.

Contrasts. Scattered acacia bushes dot the arid landscape of the "great and terrible wilderness" through which Moses once led his flock. Disheveled camels forage alongside dusty Fiats by the local Avis office. A barefoot Bedouin, bronzed by the sun and draped in flowing black robes, strides gracefully toward his cooking fire. Without breaking step, he turns to glance at the fair-skinned, bare-breasted German tourists slipping into scuba gear at the diving center. "Red Sea Divers," the proprietor's T-shirt proclaims, "the best way into the sea since Moses." Up and down this ancient coast, luxury hotels and diving centers sprawl over deserted Bedouin encampments.

Desert and sea. Ethereal and vivid. Biblical and contemporary. The Sinai has always been a land of discordant images. Yet even greater contrasts in color and life await beneath the water's surface just a few meters from shore. There, clinging to the edge of the Sinai's underwater cliffs, thrive lush aquatic gardens of which Moses' grumbling hosts, Pharaoh's sleek war galleons, and Solomon's famed trading fleets knew nothing. For this is the realm of the coral reef, an enchanted aquatic kingdom where nature has set aside all restraint, a fantasy-land whose inhabitants confound and amaze every observer with their abundance, their diversity, and their bizarre assemblages of shapes and colors.

Oddly, the water here carries little or nothing in the way of nutrients to sustain life. It is, in fact, this near-sterility that gives the gulf its clarity and sparkle; no rivers cloud it with fertile silt, and few free-floating plankton stain it green. How is it then that while the terrestrial desert stays barren, its aquatic counterpart teems with life? What conjuring trick permits this living contradiction? The answer is a masterfully engineered partnership between peculiar soft-bodied animals and microscopic plants. Working together, these unlikely bedfellows have produced a network of living rocks upon which the entire menagerie depends for food and shelter. Puzzled? Welcome to the coral reef, whose animals parade across most of the following pages.

For biologists, the reef provides an excellent context in which to explore both the unbending laws and the wild eccentricities of adaptation to life in the sea. To the trained eye and obedient lens of the photographer, the reef offers the widest assortment of technical challenges and artistic possibilities.

Most of the images in this book were captured either in the Red Sea or on others of the world's great reefs. On a planetary scale, reefs dwarf even the most impressive creations of our greatest civilizations; the Great Barrier Reef alone stretches for nearly 2,000 kilometers along the Australian coast and encompasses 21,000 cubic kilometers of limestone rock. Now add to that immensity the thousands of kilometers of reefs fringing the Red Sea and the Indian Ocean, countless coral atolls and reef-edged volcanic islands scattered across the South

Facing
Angelfish, Red Sea.

Top
Coral sea beds, Red Sea.

Above
Mouth on underside of slate-pencil
sea urchin, Red Sea.

Pacific, and the similar—though less diverse—reefs in the tropical Atlantic. From the air, many of these coral outposts evoke abstract paintings in their complexity: sheer outer walls rise to gentle slopes; spurs and grooves resemble flying buttresses on Gothic cathedrals; gracefully scalloped terraces and undulating reef flats shelter quiet, sand-bottomed, mangrove lagoons.

From a diver's normal perspective—a few meters or so from the face of the reef—the scene is wildly exotic. Hard and soft corals shaped as boulders, pillars, towers, trees, fans, nets, pigtails, and walking sticks decorate nearly every square centimeter of available surface. And like the endangered tropical forests that once formed a belt around the earth's land masses, reefs are complex, three-dimensional habitats in which the shapes and structures of dominant organisms provide homes, refuges, and nesting places for others.

For though the reef may look like solid rock, this is only an illusion; the great apparent mass of living and once-living corals is no more substantial than a sponge. Caves, tunnels, and crevices honeycomb every wall and outcropping. Some are large enough for divers to swim through six abreast, while others are so small that even the tiniest worms slither through only with difficulty. This three-dimensional complexity is one reason that hyperbole suits this habitat so well; Pacific coral reefs provide shelter for more different kinds of living things than nearly any other habitat on earth.

It takes but a little knowledge and a dash of imagination to see this labyrinth for what it actually is—a bustling alien metropolis whose offices, apartment complexes, and thoroughfares swarm with life as outlandish in form and behavior as any mythological demon. Sea serpents? Come face-to-face abruptly with a moray eel in a cave sometime, and you'll appreciate the ancients' fear of the unknown more fully. Medusa? She lives here too, in the form of a harmless, soccer-ball-sized apparition called a basket star; its Latin name, *Gorgonocephalus*, refers to a writhing tangle of snakelike arms that sweep the waters for prey.

But closer still, within a hair's breadth of the intricately patterned skin and scales of reef creatures, await images that few people—divers, photographers, or researchers—ever see. This is a world of razor-sharp spines and satin-soft fins, of gaping mouths and luminous eyes. These bits and pieces of nature's tapestry speak boldly and flamboyantly, even to the untutored eye, of beauty that matches any human artistic masterpiece. To the trained observer, however, these images whisper tales of order and chaos, of directed change and random meanderings. For these are microcosms shaped by the rules of life in the evolutionary macrocosm; bits and pieces of bodies and structures trimmed and tailored, molded and modeled to serve the needs of life in the sea.

The specific stories behind the portraits on these pages will unfold as you peruse them. Here, by way of introduction, we need only acquaint you with the rules of existence in aquatic nature and the role of light in the life of the sea. Life in the ocean—and on the reef in particular—presents conflicting customs as opposite in character as the physical appearance of the Sinai and its gulf. On one hand is the reality of "nature red in tooth and claw"; one eats or is eaten, fights for life or expires to feed another. As you will see, many features of marine animals have been shaped either to acquire some particularly tasty morsel or to avoid being made into a main course for someone else's meal.

But it is also true that the density of life on the reef is made possible only through a very different class of interactions among organisms; the sorts of partnerships biologists call symbiosis, or "living together." Some of these partnerships are visible and easily recognized as relationships of consequence. Such

is the familiar alliance between clownfish and sea anemone. Here, the fish finds shelter and a spot to hide its eggs and young amidst stinging tentacles that deter most of its enemies. But the anemone has enemies of its own—butterflyfish and angelfish who like nothing better than a meal of tender tentacle tips—that are driven away by the pugnacious resident clownfish. Life on the reef is replete with affiliations such as this one; some just as apparent, some less obvious, and some completely imperceptible to the naked eye. It is through these partnerships, the most important of which involve corals, that these oases of life flourish in stretches of ocean that are otherwise barren.

At the root of all life, however, is the energy that powers the machinery of cells, tissues, and organs. Virtually all of that energy comes to earth in a form so abundant, so dependable, and so ordinary, that we usually think little of its magic. The source of that energy, of course, is the sun. Were we not so blasé about sunlight's effects, we would surely find them suitable subjects for a fairy tale. No god of ancient Rome, after all, had powers more magical than those of the invisible rays that bathe our planet. In their glow, nonliving matter comes to life, as light energy powers the growth of plants, algae, and the few other living things that can collect and harness it. And although we seldom think of it that way, those rays indirectly fuel the life processes of animals, both those that eat plants directly and those that eat each other.

At the same time (and almost incidentally) sunlight energizes one of the most important channels of communication between animals and the world around them: the sense of sight. That we see by the grace of sunlight is evident to all. But we consider less often that the spectacularly colored structures of animals and plants display their hues only because of the light that strikes them and the eyes that see them. For it is by reflecting and absorbing light, by playing tricks with energy, that the bodies of living things speak to the eyes of friends and foes alike. Whether that language speaks the truth or whether it lies, whether it broadcasts to all or whispers only to those with reason to listen, depends on the history and habits of the animal in question.

You will find, as you browse through the rest of this book, that all these photographs can speak to you in several ways. Each image represents a specific example of the themes of life beneath the sea: partnership and confrontation in lifestyle, deceit and integrity in communication, the dependence of marine life on the unusual nature of light in the sea, and the relationship between form and function. On another level, all the photographs underscore the incidental beauty to the human eye of creatures whose appearance has evolved only to suit one another. And in the most universal sense, these intimate glimpses of faces, fins, and forms testify to both the boldness and the subtlety of Nature's brush in painting her subjects.

Facing
Butterflyfish, Red Sea.

Pages 26-27
Sea anemones, Palau.

Pages 28-29
Symbiotic shrimps on a nudibranch,
Red Sea.

Pages 30-31
Aerial view of a reef, Palau.

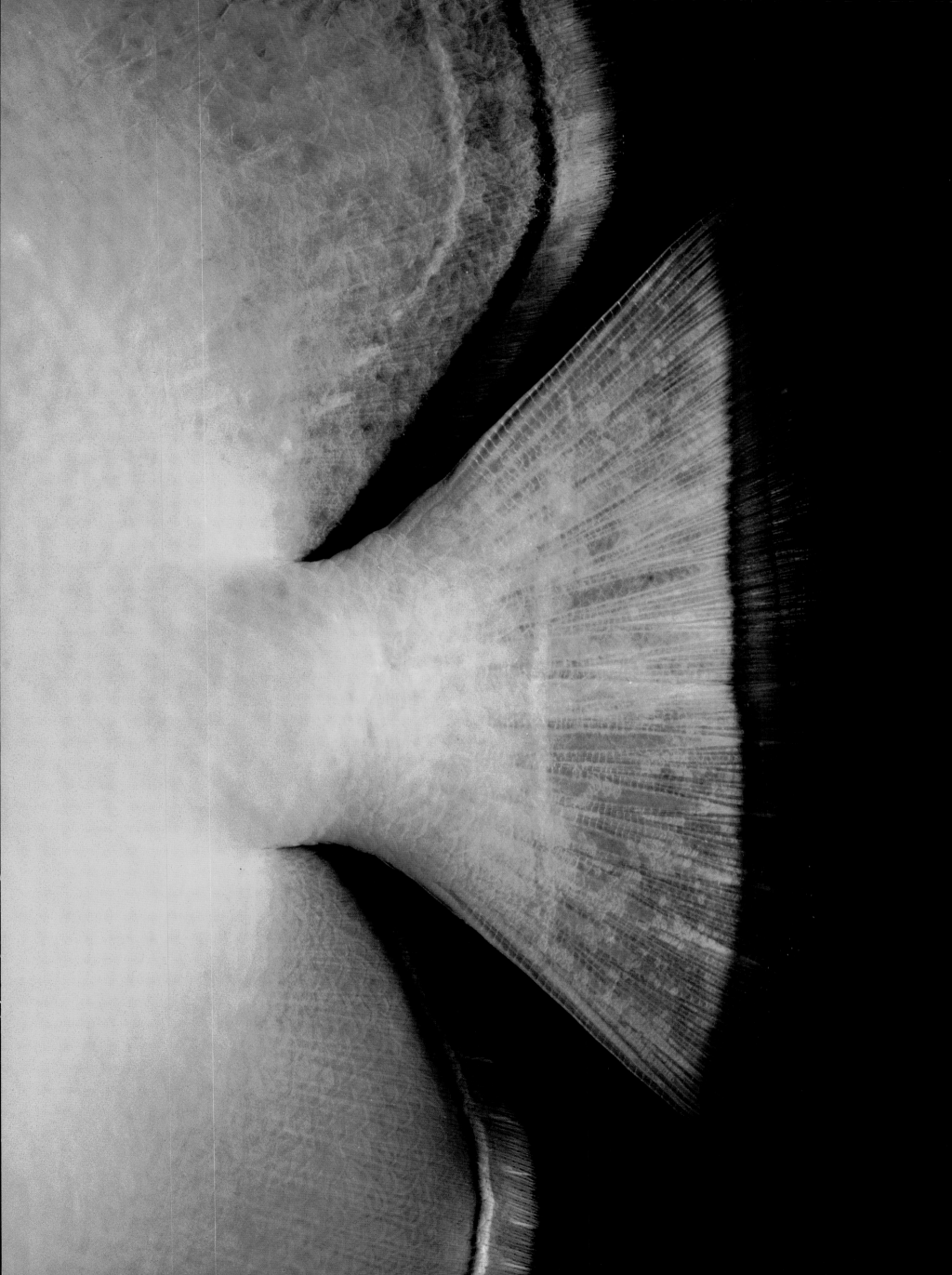

LIVING·RAINBOWS

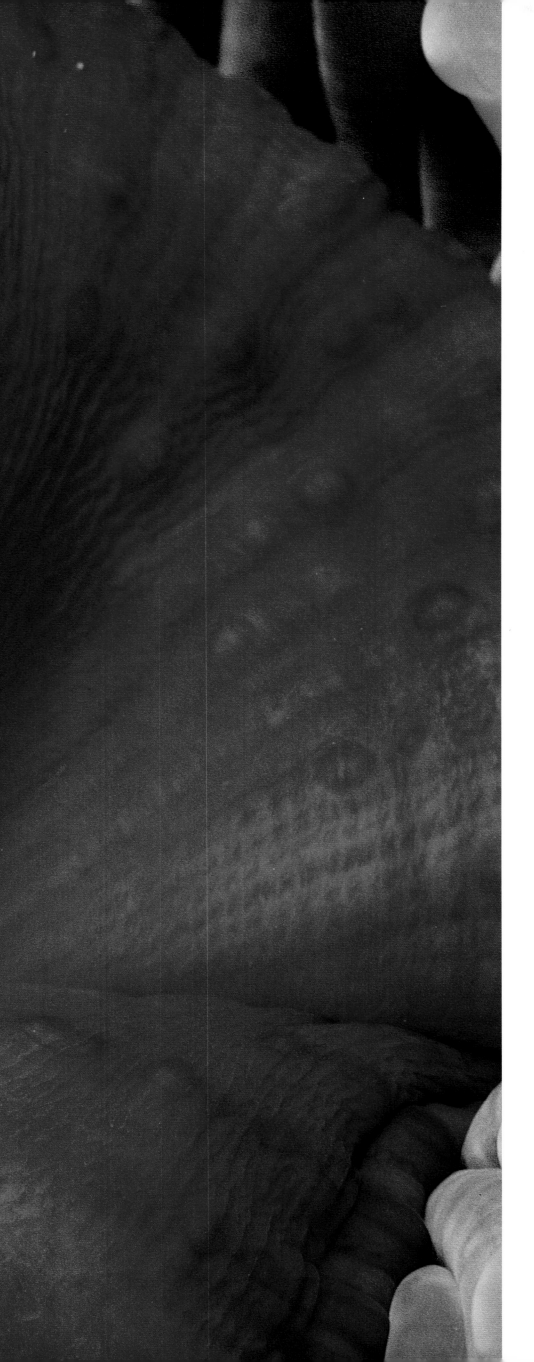

No matter what you read, nothing can prepare you for your first dive on a coral reef. Fishes and other animals covered with extravagant colors and intricate patterns lurk in the shadows, skitter across piles of rubble, and wheel amidst coral formations. Here you can find virtually every hue in the rainbow, displayed separately and in combinations that fashion designers would envy. It is easy to see why early European explorers described the reef's inhabitants as adorned with "polished scales of gold, encrusting lapis lazuli, rubies, sapphires, emeralds, and amethysts."

But where do these living rainbows come from? How are their colors formed? Our answers invoke both the knowledge of scientists and the insights of artists, for the laws of physics dictate that both animals in nature and artists in studios must produce colors in essentially the same ways. The only real difference between human works of art and patterns on the skins of marine creatures is that the former are created from paints and canvas, while the latter are painted literally with "living colors"—by skin cells that manipulate light like splashes of color in Impressionist paintings.

At the root of it all are the phenomena of light and color. The visible part of sunlight, or "white" light, contains a range of wavelengths from violet at one end to deep red at the other. Things that look white, such as an egret's plume or the glistening peaks of the Alps in winter, reflect a mixture of all these wavelengths. The velvety-black fur of a panther, on the other hand, absorbs most visible light, sending very little to our eyes. But brightly colored flowers, butterflies, and fishes carry pigments that absorb light of some wavelengths and reflect others. Ripe strawberries, for example, reflect red light, while absorbing light green and blue.

As human artists struggled to re-create the countless colors they saw around them in nature, they learned to reproduce nearly any hue by combining three primaries—red, yellow, and blue—with the correct amounts of black and white. To begin, many (though not all) artists set up their canvases with a coat of whitewash. This reflects back any light that hits it, and so adds brilliance to the composition. On top of the white layer, they apply paints either separately or in combination—in mixtures, in layers, or as fine lines or spots of paint next to one another. The simplest mixtures are familiar to anyone who's ever dabbled with finger paints or crayons: red and yellow together produce orange, blue and yellow make green, and red and blue create purple. Other tricks are less obvious, but equally important: blending black into lemon yellow transforms it to green; adding black to orange takes it down to a warm brown; adding blue and white to red creates pink.

The first humans to think of whitewashing and color mixing in this way probably thought they were clever. Yet animals in the sea had beaten them to it by millions of years. Many reflect light like microscopic mirrors. In fish skins, these plates are arranged to reflect light evenly in all directions, producing the dazzling whites we see in many butterflyfishes.

Pages 32-33
Parrotfish, Red Sea.

Left
Symbiotic clownfish on an anemone, Red Sea.

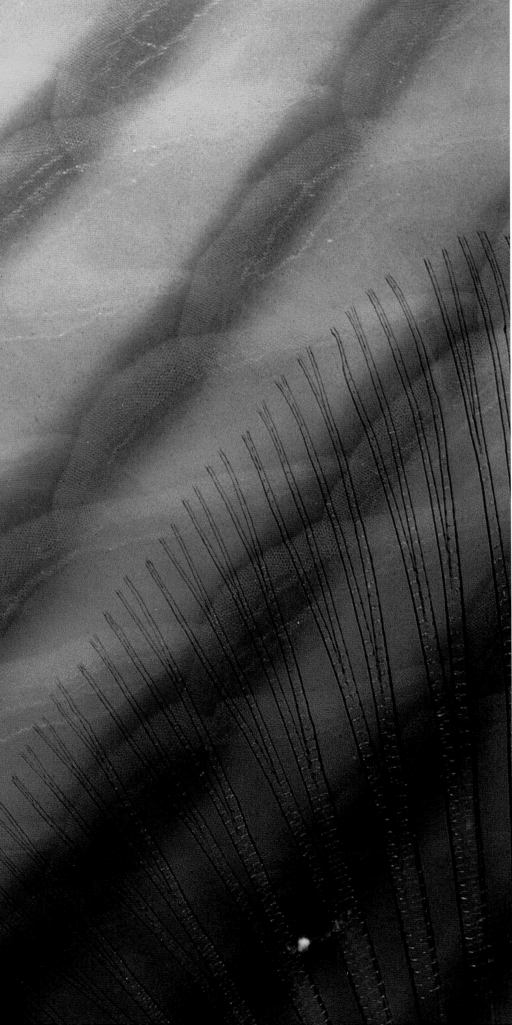

Atop this sparkling background, animals, like human painters, layer other colors. For the dark end of their palettes, they draw upon pigments that range in color from brown to black. Using these dark pigments and their "whitewash" alone, fishes create black-on-white patterns that predated the stark contrasts of contemporary haute couture by eons. Then, by adding extra cells carrying yellow, orange, and red pigments, the more flamboyant angelfishes and parrotfishes complete their ensemble. Many marine animals procure their colors much as human artists did for centuries, by collecting "vegetable dyes"—pigments produced by plants. Of course, fishes and other sea dwellers don't boil algae to extract the pigments. Instead, they eat the plants, remove the pigments during digestion, modify them slightly if necessary, and finally funnel them into appropriate cells in their skin. The net result is a sea filled, quite literally, with every color in the rainbow, and several hues the skies have never held. As human observers, naturally, we see this rainbow from our species' own particular perspective. Common sense suggests that the animals themselves might see things differently; perhaps they see a completely different rainbow or none at all. Is there any way to tell?

The answer is both yes and no. There are some colored pigments in the ocean, particularly in the deep sea and among strictly nighttime animals, that have little or no function as colors per se. (The deep-red hemoglobin that colors the blood of some deep-sea worms comes to mind; the rich red color of that oxygen-carrying compound is strictly coincidental in that case.) There are other colors, including those of many soft corals, that seem incidental but may have functions of which we know nothing. We have as yet no adequate test to determine whether or not animals see things as we do in such cases.

On the other hand, we know from research into the workings of fish eyes and brains that many reef fishes do see a world filled with colors. We also know that these animals do not see the world precisely as we do. Some species see many more colors in the rainbow than we can, and many see well in either ultraviolet or infrared parts of the spectrum where human eyes perform dismally. Yet, beyond these particulars, the networks of nerve cells in the eyes and brains of both fishes and humans process visual information in similar ways. The result seems to be a remarkable overlap between the visual world of at least some reef animals and that of our own species.

How can we make that judgment? Colors in the sea evolved, after all, because of their appearance to their bearers and other marine animals—not because they look good to us. That camouflage coloration evolved to fool fishes fools our eyes as well tells us that the world as fishes see it must have at least some resemblance to the world we see. We as observers should be grateful for that incidental perceptual overlap. For if these animals used vision but little, or if they viewed the world through radically different sorts of eyes, much of the beauty and wonder of the reef might be invisible to humankind.

Left
Butterflyfish, Red Sea.

Left
Brain coral, Red Sea.

Right
Daisy coral polyps, Red Sea.

Left
Black-spotted pufferfish, Red Sea.

Above
Sea anemones, Red Sea.

Pages 40-41
Sea anemones, Red Sea.

Above
Anemone mouth, Point Lobos,
California.

Facing
Big-eye scad, Red Sea.

Pages 44-45
Dorsal fin of a soldierfish, Red Sea.

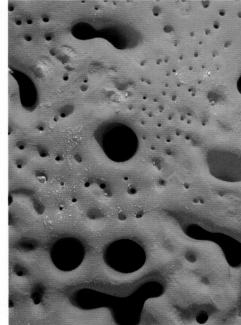

Left
Caudal fin of a jewel grouper, Red Sea.

Top
Eggs beneath a lobster's tail, Ustica Island, Mediterranean Sea.

Above
Sponge, Caribbean Sea.

Pages 48-49
Armored tail of a surgeonfish, Red Sea.

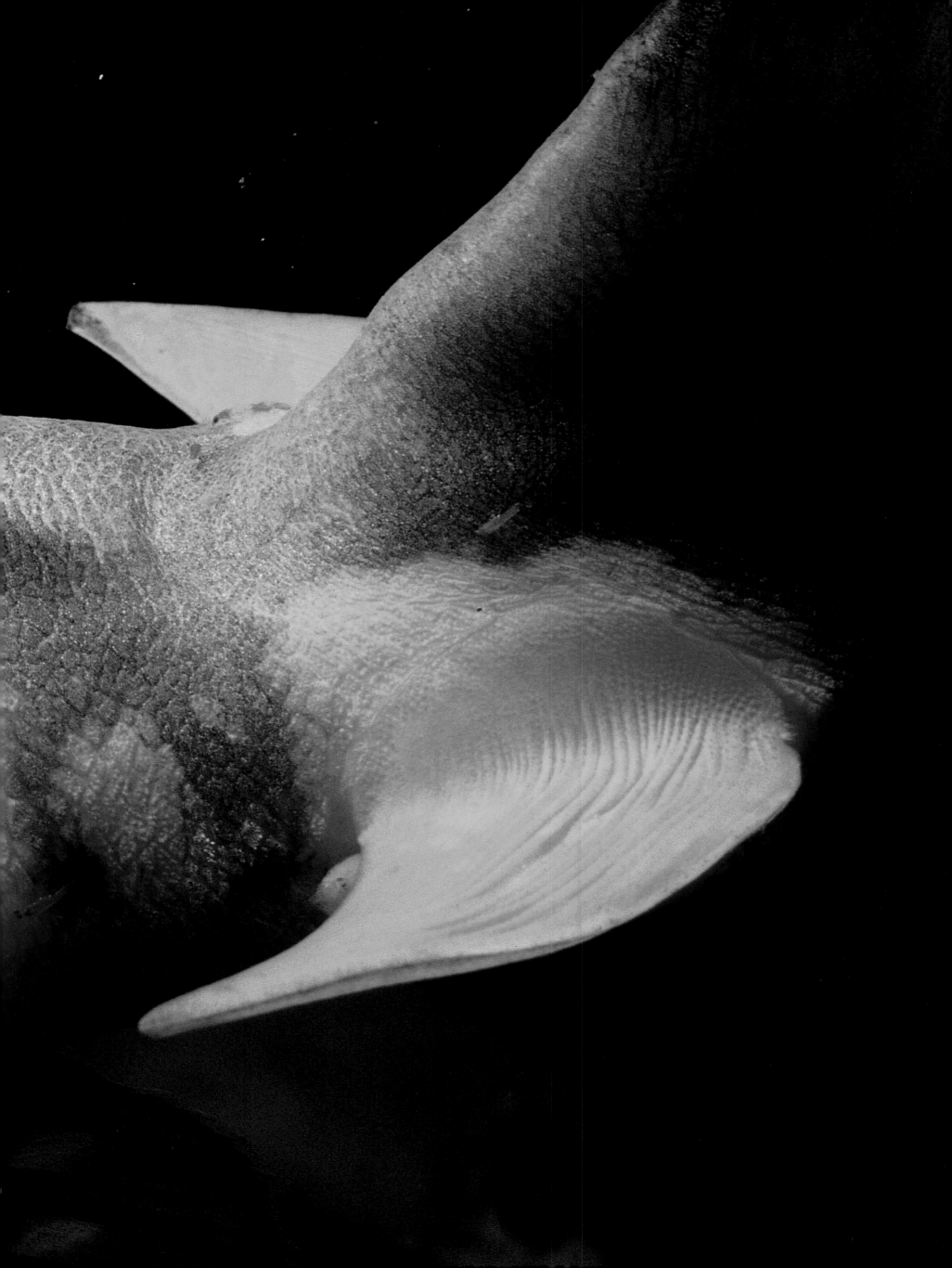

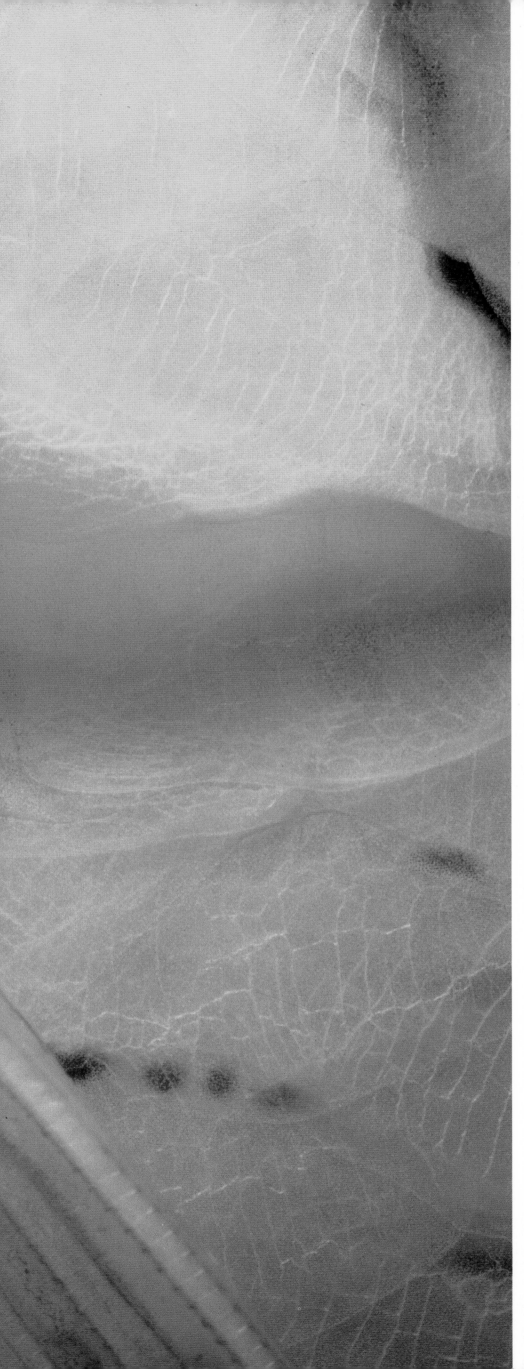

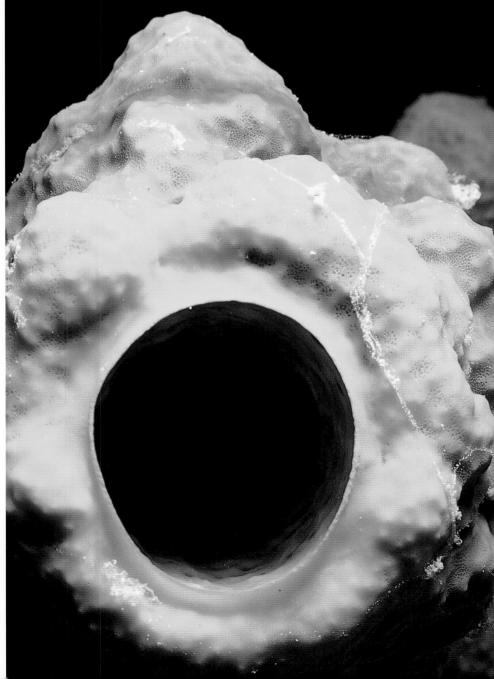

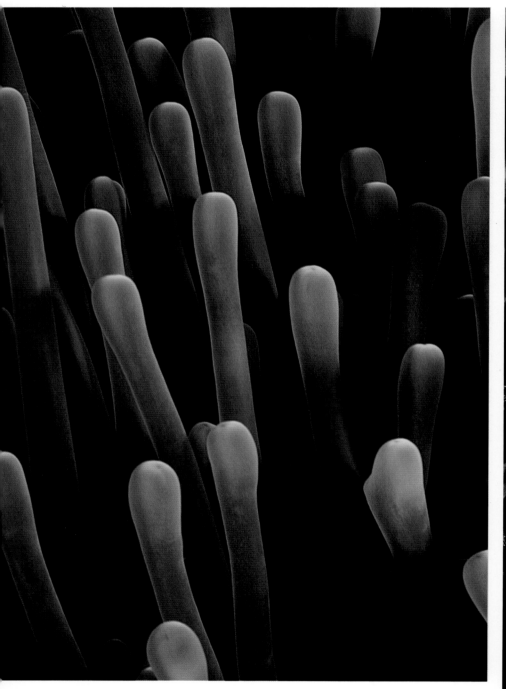

Pages 50-51
Sling-jaw wrasse, Red Sea.

Page 51, top
Yellow jack, Red Sea.

Page 51, bottom
Cylindrical sponge, Cayman
Islands.

Above
Anemones, Palau.

Right
Crinoid, Palau.

Left
Corals and algae in symbiosis, Palau.

Right
Coral polyps, Caribbean Sea.

Page 56
Angelfish, Red Sea.

Page 57
Caudal fin of a parrotfish, Red Sea.

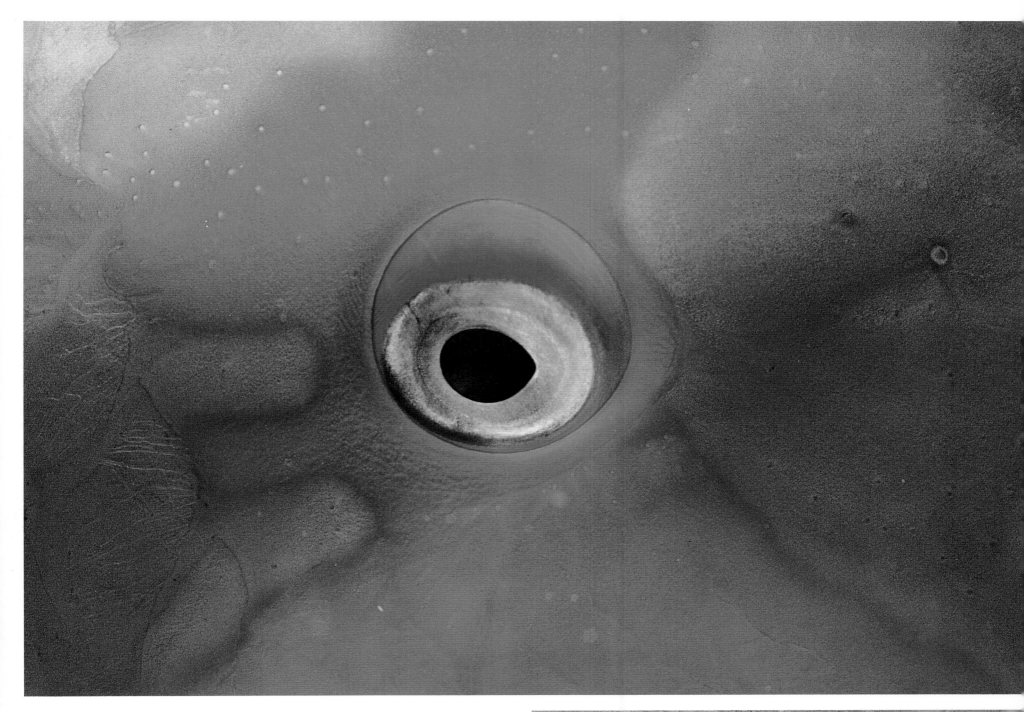

Facing
Caudal fin of a parrotfish, Red Sea.

Above and right
Parrotfish, Red Sea.

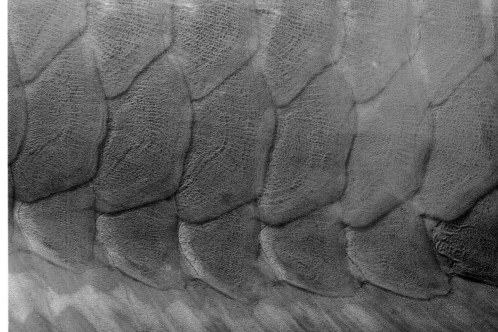

Below
Crinoid arm, Red Sea.

Bottom
Tube sponge, Caribbean Sea.

Right
Sea urchin mouth, Ustica Island, Mediterranean Sea.

Pages 60-61
Mushroom coral, Palau.

Pages 62-63
Soft coral, Red Sea.

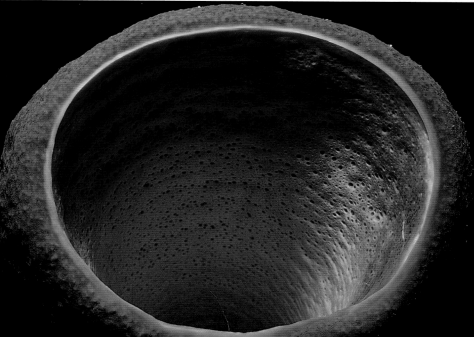

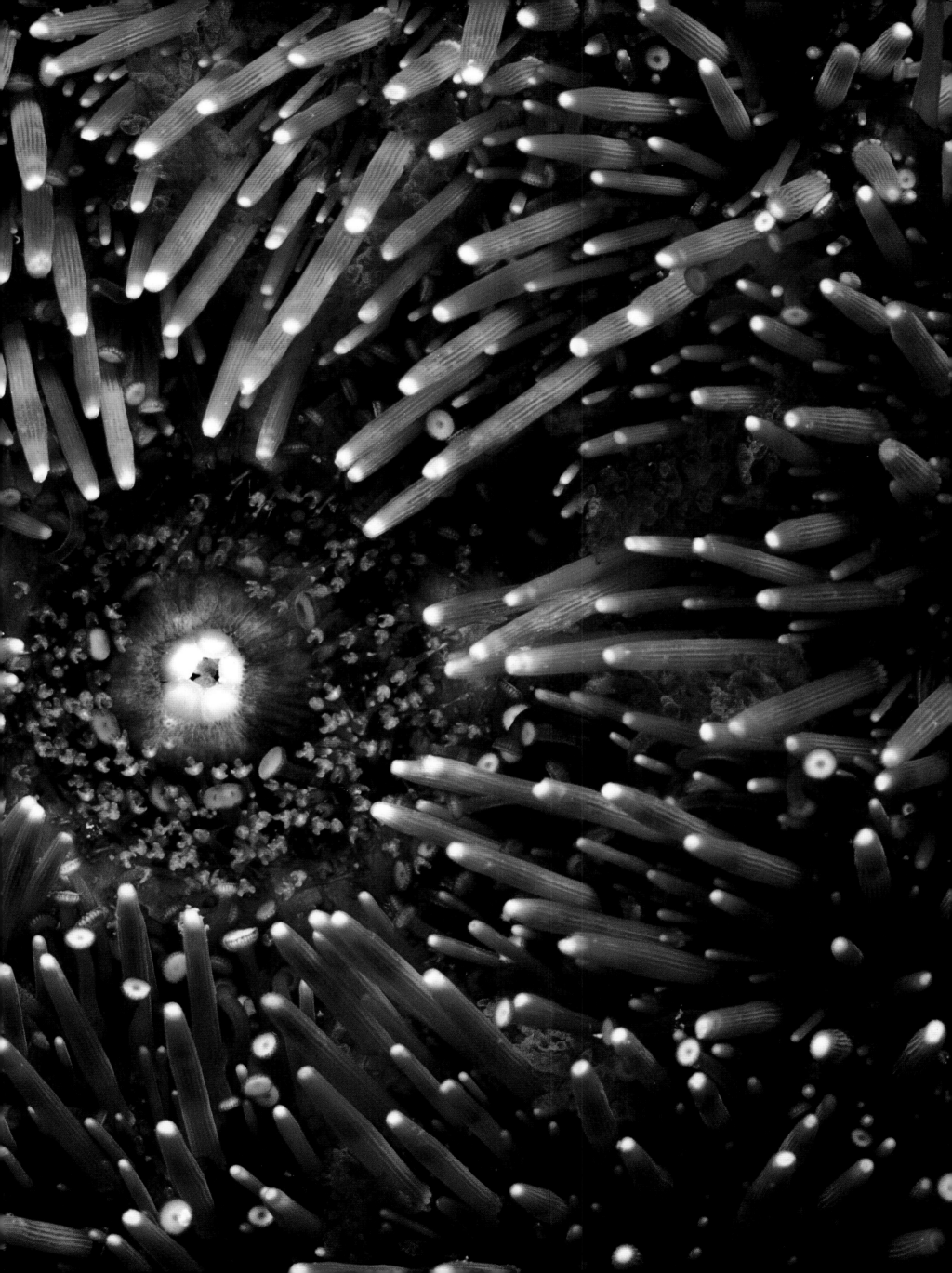

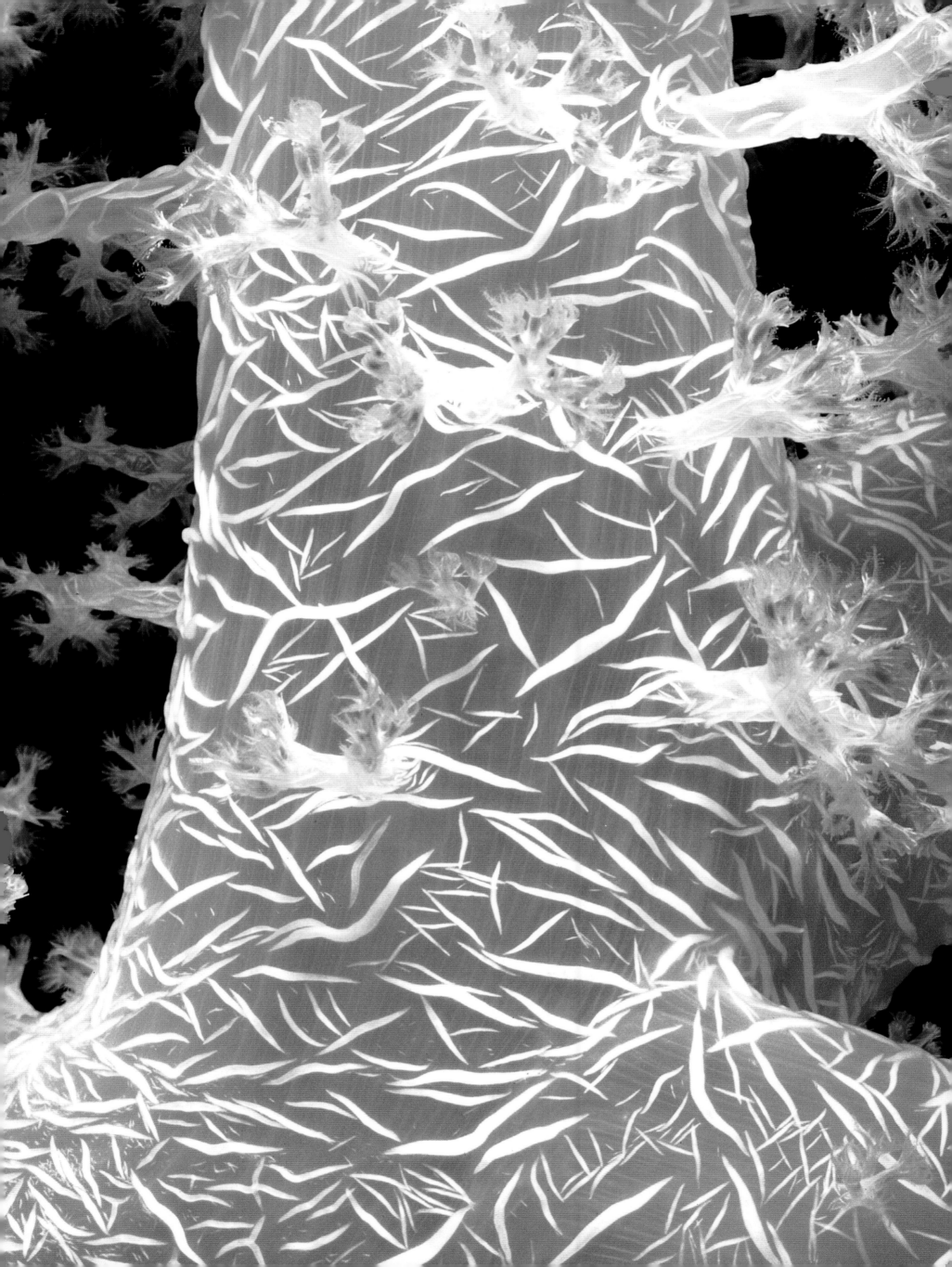

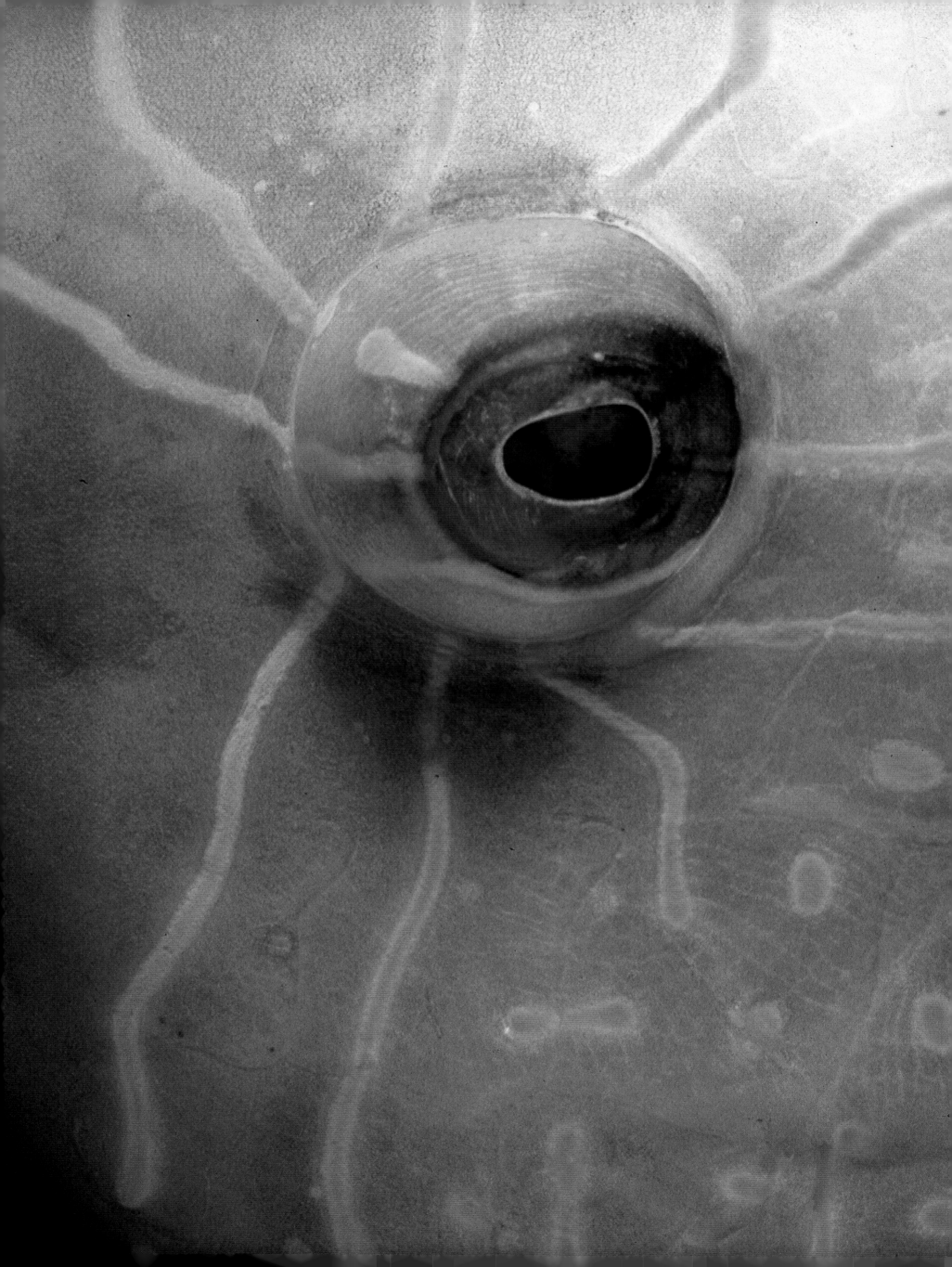

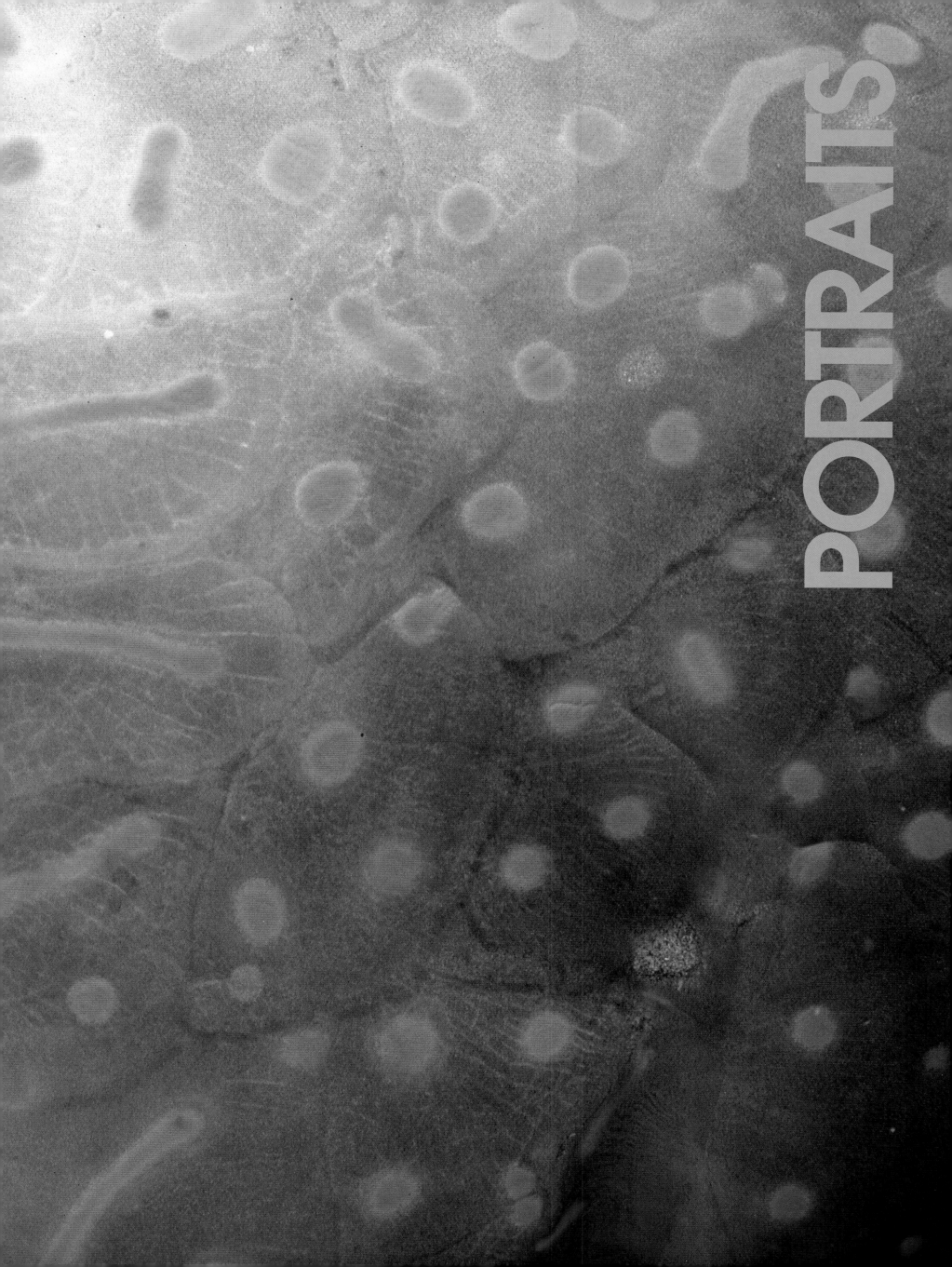

PORTRAITS

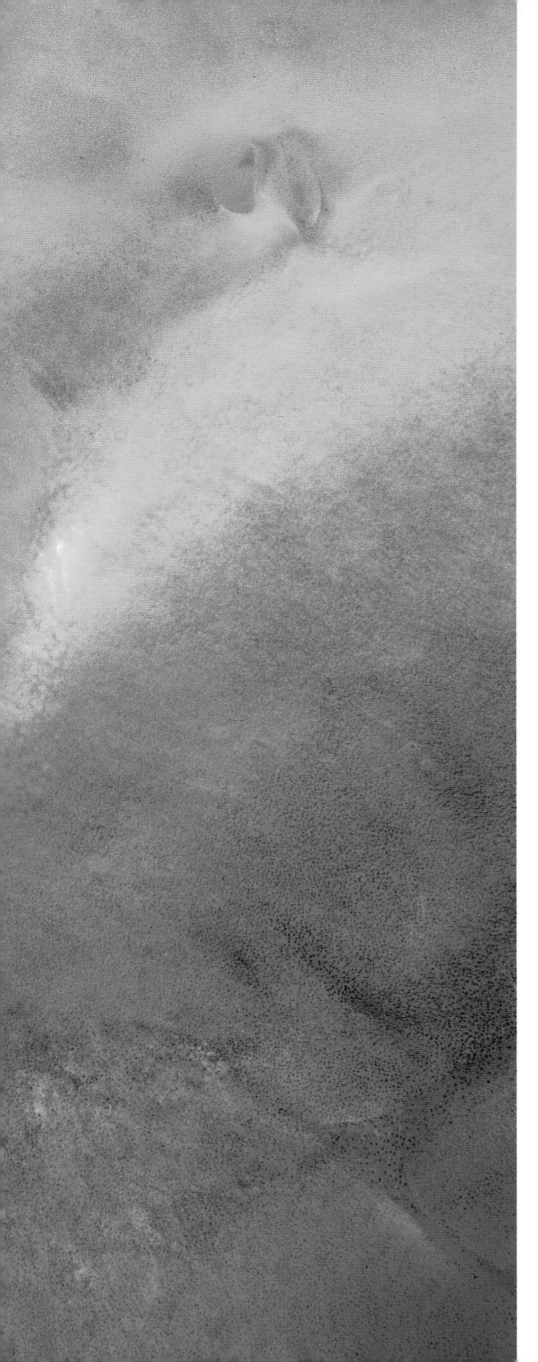

Snouts. Outlandish assemblages of scales, jaws, and teeth worthy of creatures from outer space nuzzle inquisitively through soft-coral thickets, lurk amidst rocks covered with multicolored algae, and cruise over featureless mud flats.

Eyes. Small and beadlike or huge and staring, prominently displayed or hidden by colors of skin and scales, they scan the sea for friends, foes, and food.

Together, mouths and eyes give fishes the "personalities" we're likely to read into their faces; bold or shy, harmless or menacing, beautiful, coy, otherworldly, or just plain odd. But to those of us who know fishes well, their faces do more than imply temperament; they bear witness to fascinating exploits in survival and adaptation. For each fish face, shaped through the eons by natural selection, speaks volumes about its bearer's history and place in the complex world of life in the sea.

Take the scarlet bigeyes and squirrelfishes, for example. Their ancestors, among the first of the modern fishes to challenge the older sharks for control of the sea, patrolled the reef while dinosaurs ruled the land. Like many early fishes, these ancient species used simple, open-and-shut mouths to snap up small fishes, shrimp, and crabs. Over the years, the mouths of fishes in these families stayed pretty much the same, even as other, more recently evolved groups developed increasingly complicated and adaptable ways of feeding.

In part because of competition from those better-equipped newcomers, bigeyes and squirrelfishes were forced out of the limelight and into the shadows; now they spend their days lurking in darkened reef caves and crevices. Only under cover of night do they venture out into open water. Obliged to find food and avoid enemies in dim light, these species evolved huge, staring eyes as sensitive as those of a cat. These eyes, however, are not as adaptable as those of our fellow mammals. Lacking irises to control the amount of light that enters, nighttime fishes' eyes are easily "overexposed" in bright sun, leaving the animals practically blind. Even under the best of circumstances, these fishes cannot see the colors that surround them on the reef; their world consists of shadowy images in black, white, and shades of gray.

To understand why these older, more primitive fishes were shouldered aside, we need look no farther than the faces of more modern species; there we can see the sorts of new and improved "equipment" that made these animals stronger competitors in the struggle for food and space on the reef. Groupers, for example, hunt not in real darkness, but at twilight, when their gaping, fang-filled jaws swallow in a single gulp any fishes and crabs they manage to catch off-guard. That's not an easy thing to do, because the prey animals can see almost (though not quite) as well as groupers can at dusk. To help stack the odds in the grouper's favor, its mouth does some tricks that the bigeyes never mastered.

Look closely at the grouper's large upper "lip"; it is actually a movable bone, connected to the head by flexible ligaments and ample folds of

Pages 64-65
Wrasse, Red Sea.

Left
Yellow jack eye, Red Sea.

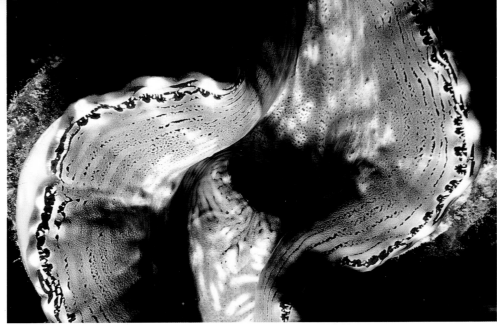

Left
Tridacna clam, Red Sea.

Below
Red-lipped batfish, Galápagos.

Right
Parrotfish, Red Sea.

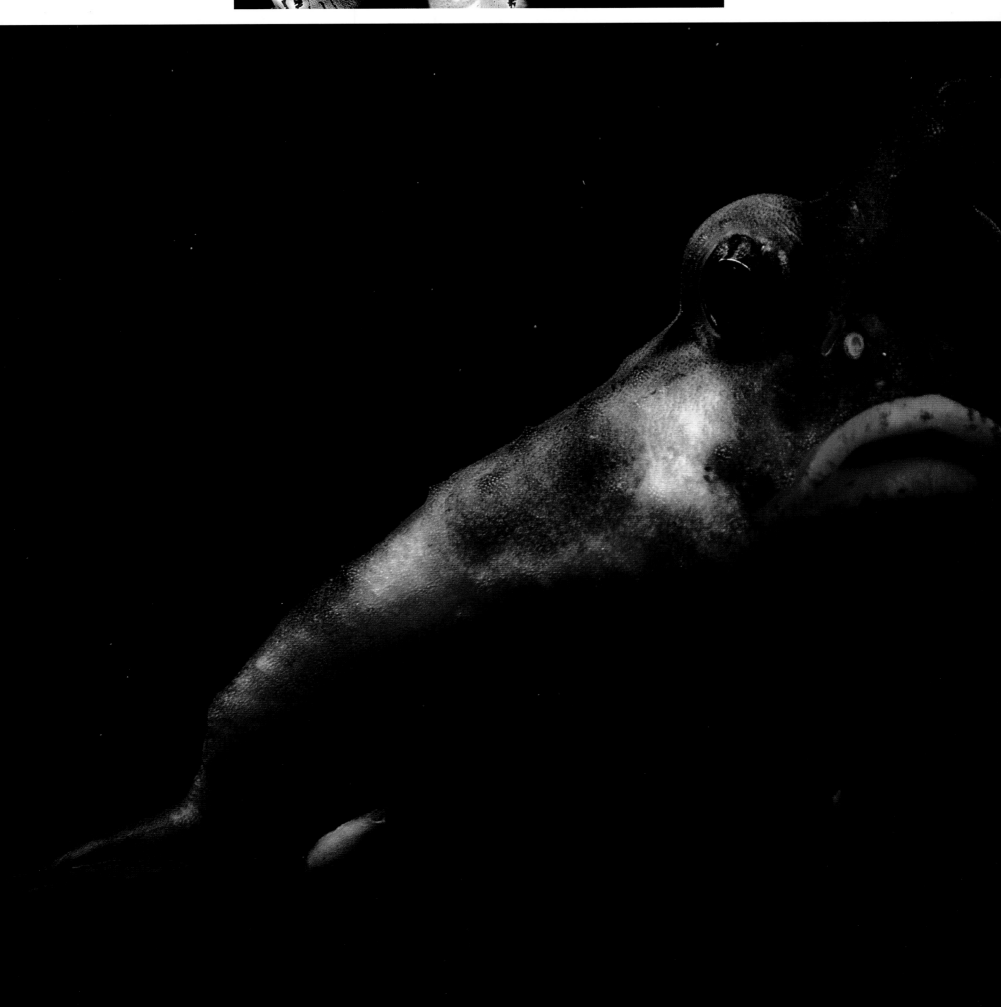

skin. Behind it is another bone, less mobile, but hinged in front and free to swing down in back. When the fish drops its lower jaw, that first upper bone slides down and out, while the bone behind it swings down. Together, they reshape the mouth into a tube that jumps outward, ahead of the fish, to suck in prey like a vacuum cleaner. This whole sequence happens faster than the blink of an eye; to see it, researchers use ultra-high-speed cameras that divide each second into scores of separate images.

Many fishes carry this trick even farther, as other closeup portraits show. The bright yellow sling-jaw wrasse carries what appears to be a great deal of "extra" skin in folds behind its jaws. But all that slack has its purpose; when this fish opens its mouth, a complicated assortment of levers, joints, and pulleys makes those jaws virtually explode into a long, narrow tube that engulfs small prey before they know what hits them. (To give you an idea of just how far this jaw protrudes, on a scale in which the wrasse's face fills an entire page in this book, its extended mouth would reach halfway across a facing page.)

Of course, snatching prey that can scamper away is not the only way to make a living in the sea. Many fishes favor other sorts of food, and the design of each species' mouth reflects its way of life. The tiny mouths of angelfishes and butterflyfishes don't pop out very far when opened, and carry arrays of slender teeth that function as combs, rather than as fangs. These mouths enable their bearers to nibble on corals, sponges, and the algae that encrust most sunlit surfaces on the reef. Parrotfish use their powerful, birdlike beaks to chew through coral rock with the ease of children nibbling on soda crackers. Naturally, the fish don't digest the rock dust; but along with all that finely ground debris they pick up the bits and pieces of algae that form the mainstay of their diets.

Some species, such as puffers and certain skates, feed on crabs, snails, clams, small mussels, and other animals that depend on thick shells for protection. Puffers bite through that armor-plate with sharp teeth and powerful jaws, while many skates crush their victims with rows of short, blunt teeth.

Scorpionfish, which ambush their quarry, have eyes and mouths that are shaped and colored, not only to function well, but to disappear whenever possible. Flatfishes, more concerned with hiding from predators than from prey, use similar tricks to blend in with the drab, yet textured, sandy and muddy bottoms on which they rest.

Closeup views of fish faces reveal other secrets as well. Look again at a bigeye, and notice a row of small pores between eyes and upper jaw. These holes open into a network of sensory canals that run, out of sight, through and beneath the scales. Often described as the aquatic equivalents of ears, these canals actually function quite differently from our own organs of hearing. By detecting tiny vibrations in the water, they allow fishes to "hear" predators and prey moving nearby.

As you can see, the variations evolution has played on fish faces are as diverse as fishes themselves. On that diversity, scientists can wonder, and all of us can continue to explore with delight.

Pages 72-73
Grouper, Red Sea.

Pages 74-75
Needlefish, Red Sea.

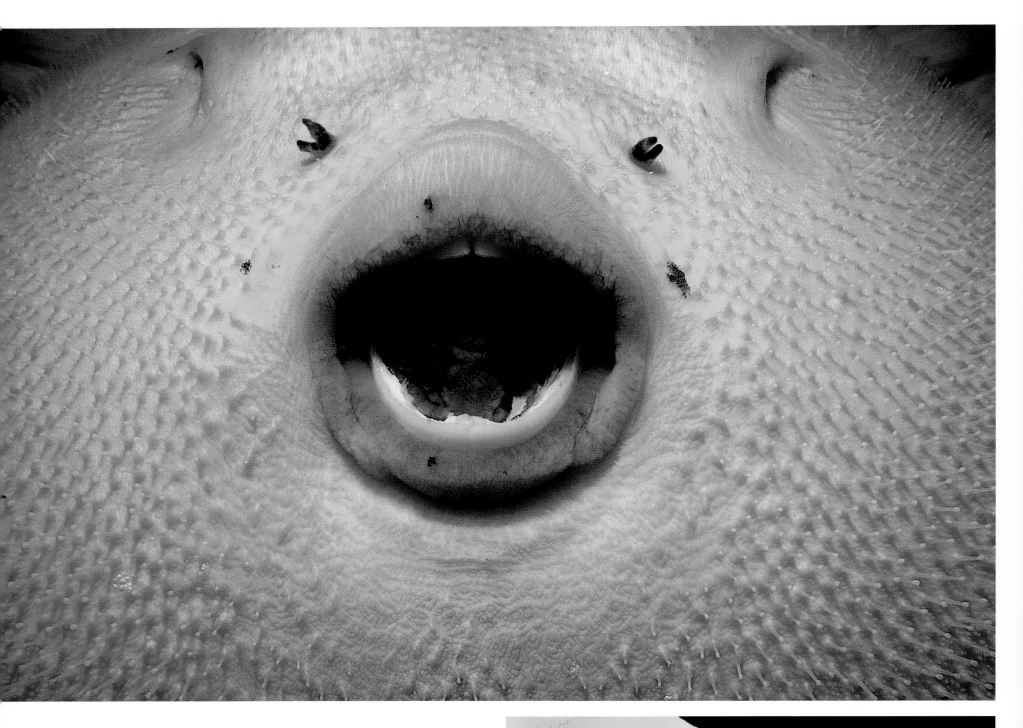

Above
Pufferfish mouth, Galápagos.

Right
Surgeonfish, Red Sea.

Facing
Wrasse, Red Sea.

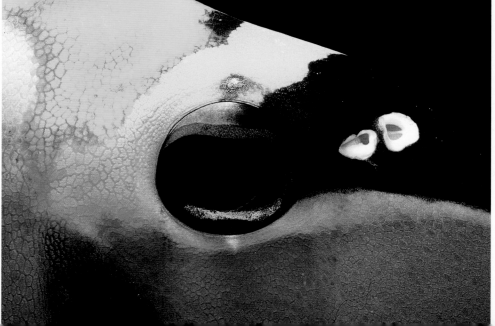

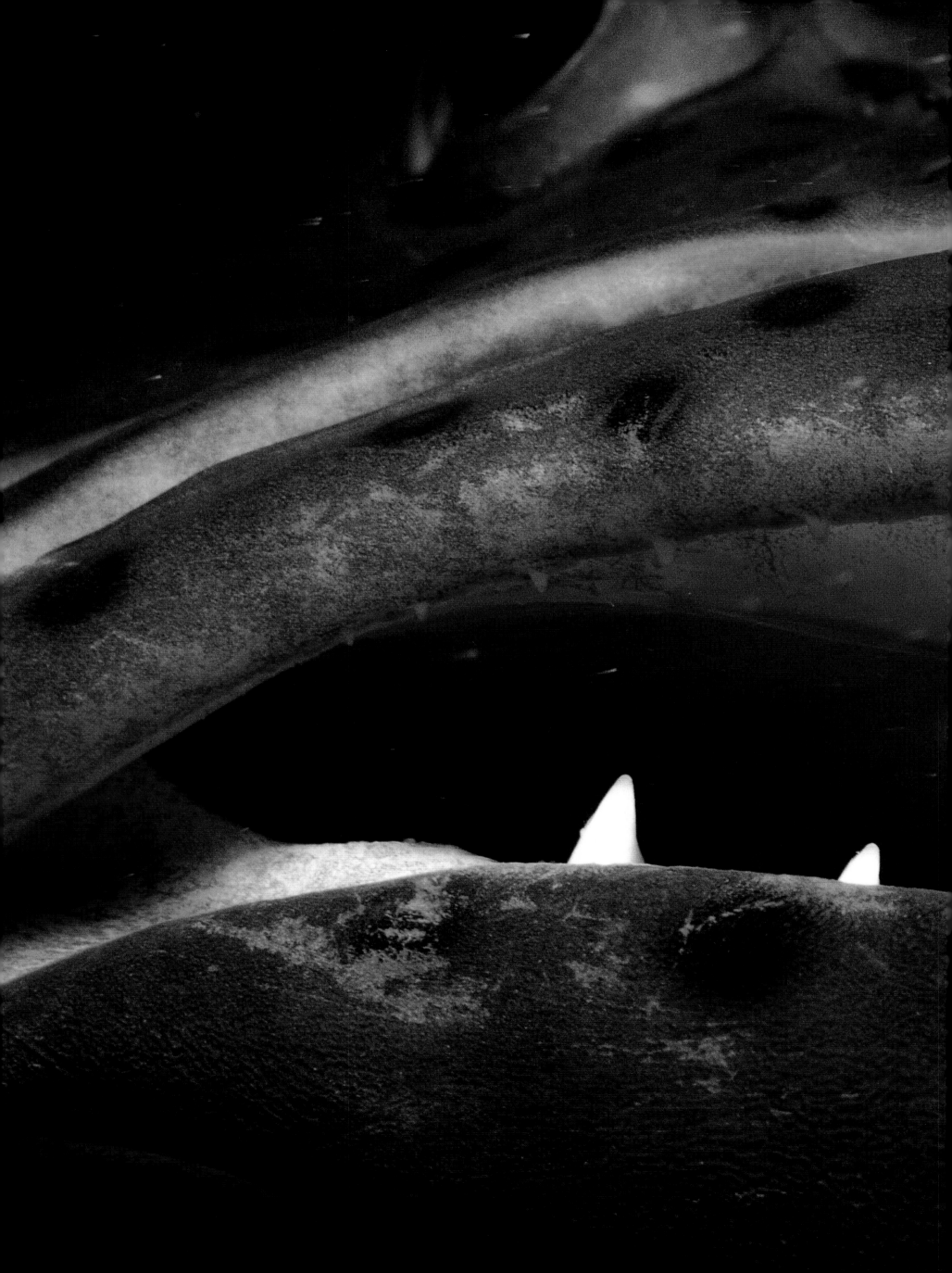

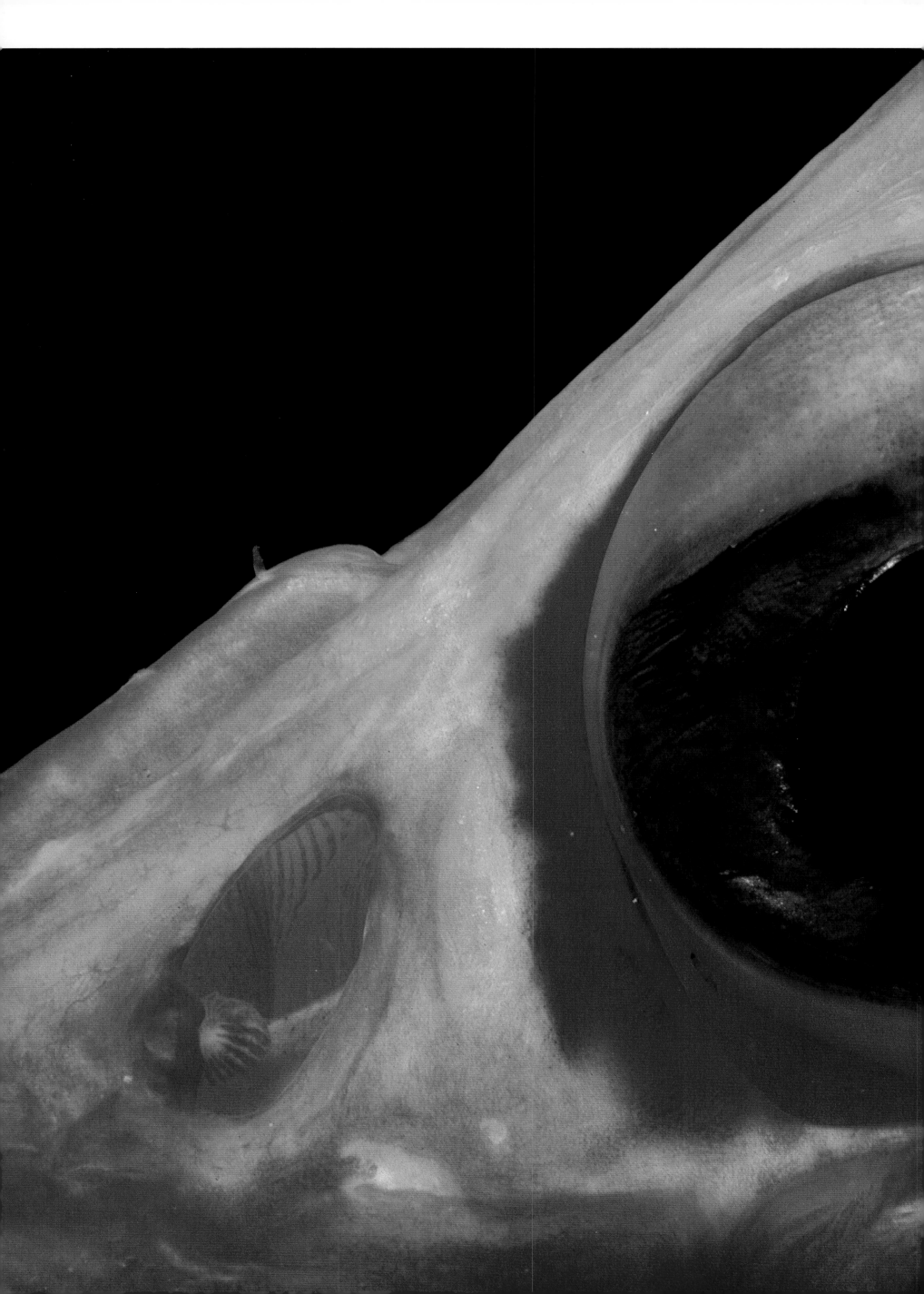

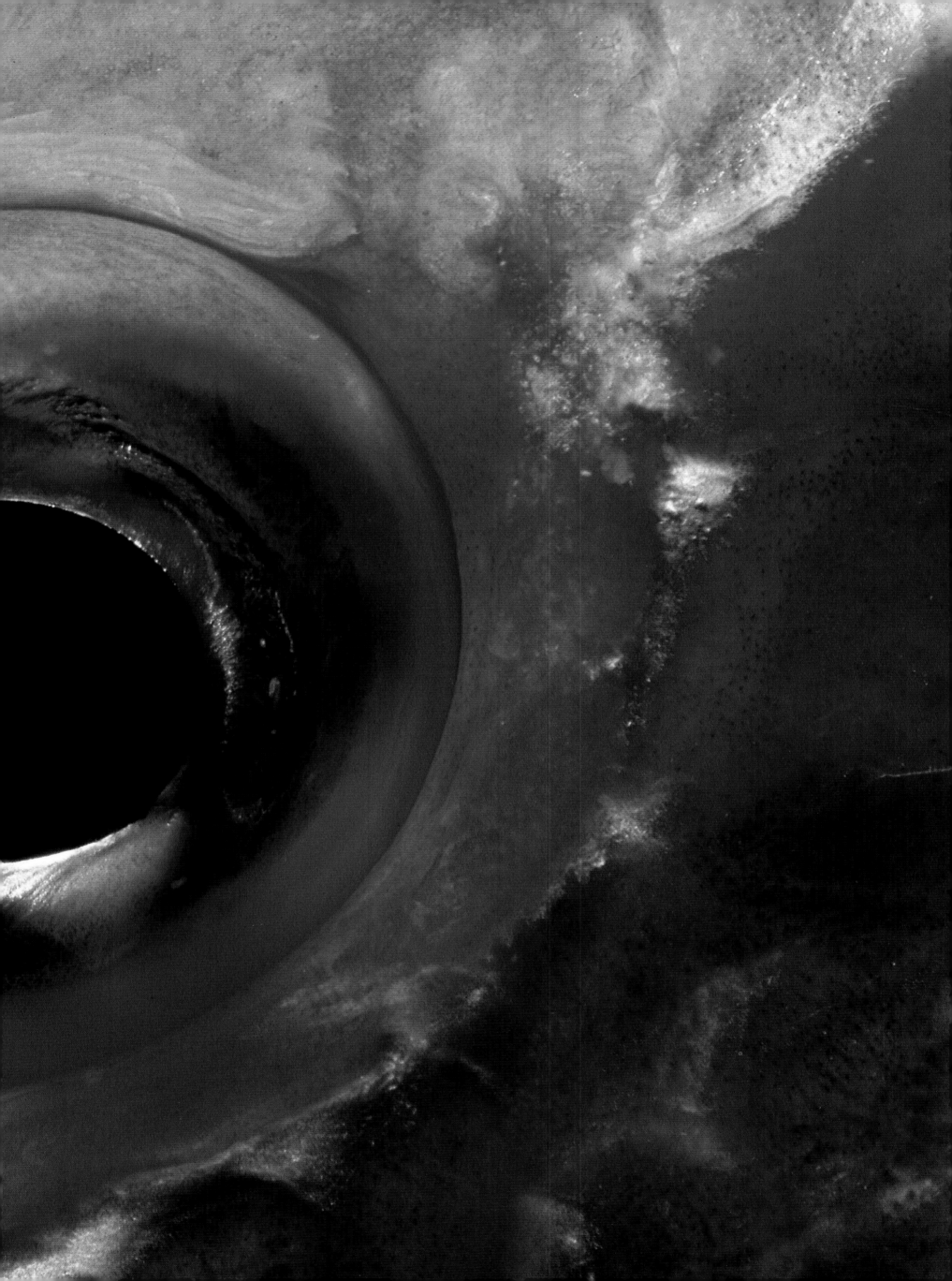

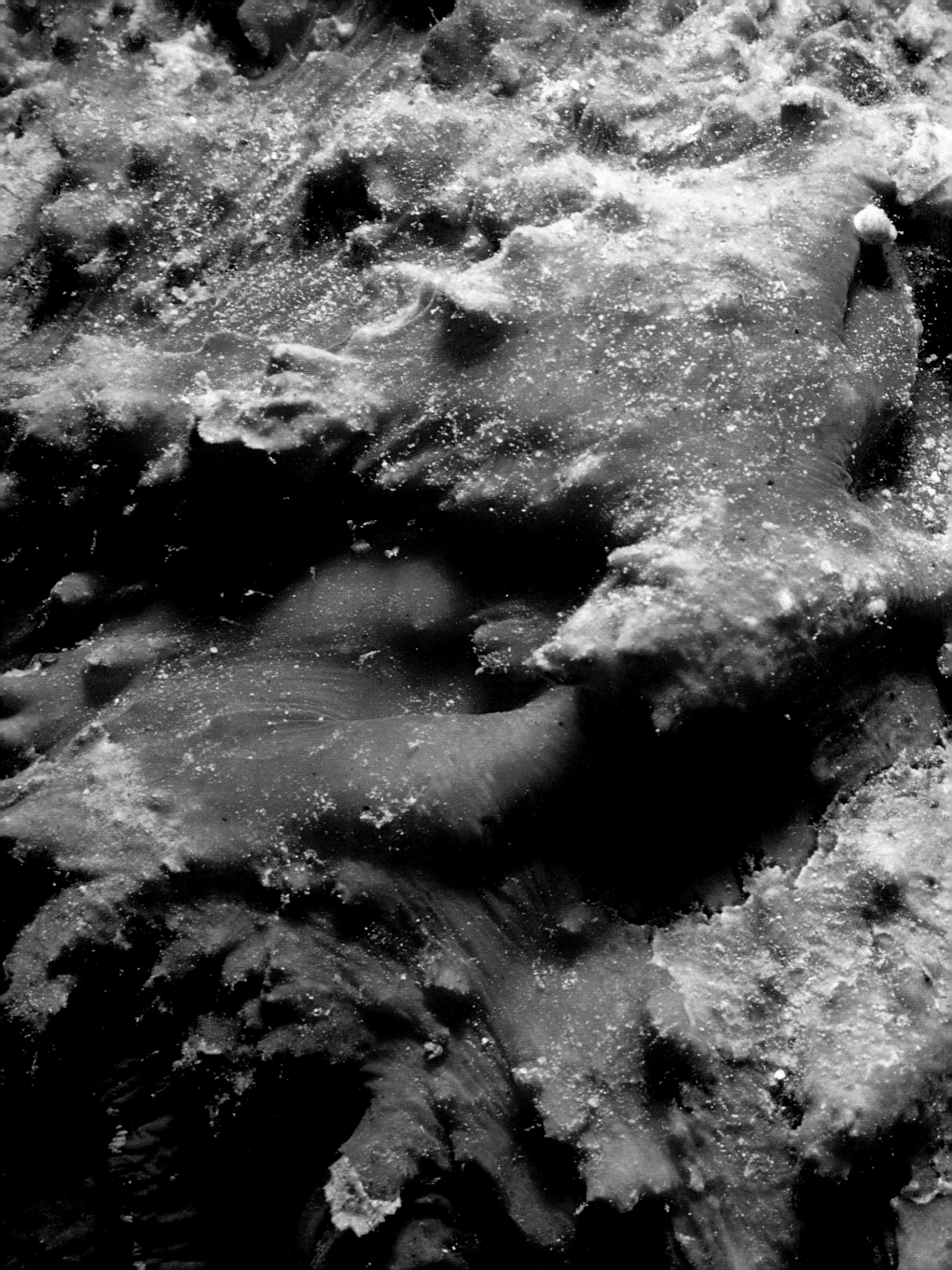

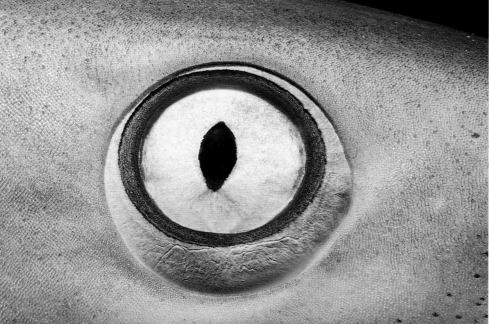

Left
Lemon shark, Bahamas.

Above
Yellowtail flounder, Gloucester,
Massachusetts.

Above
Crab, Red Sea.

Right
Skate, Monhegan Island, Maine.

Left
Hermit crab, Ustica Island,
Mediterranean Sea.

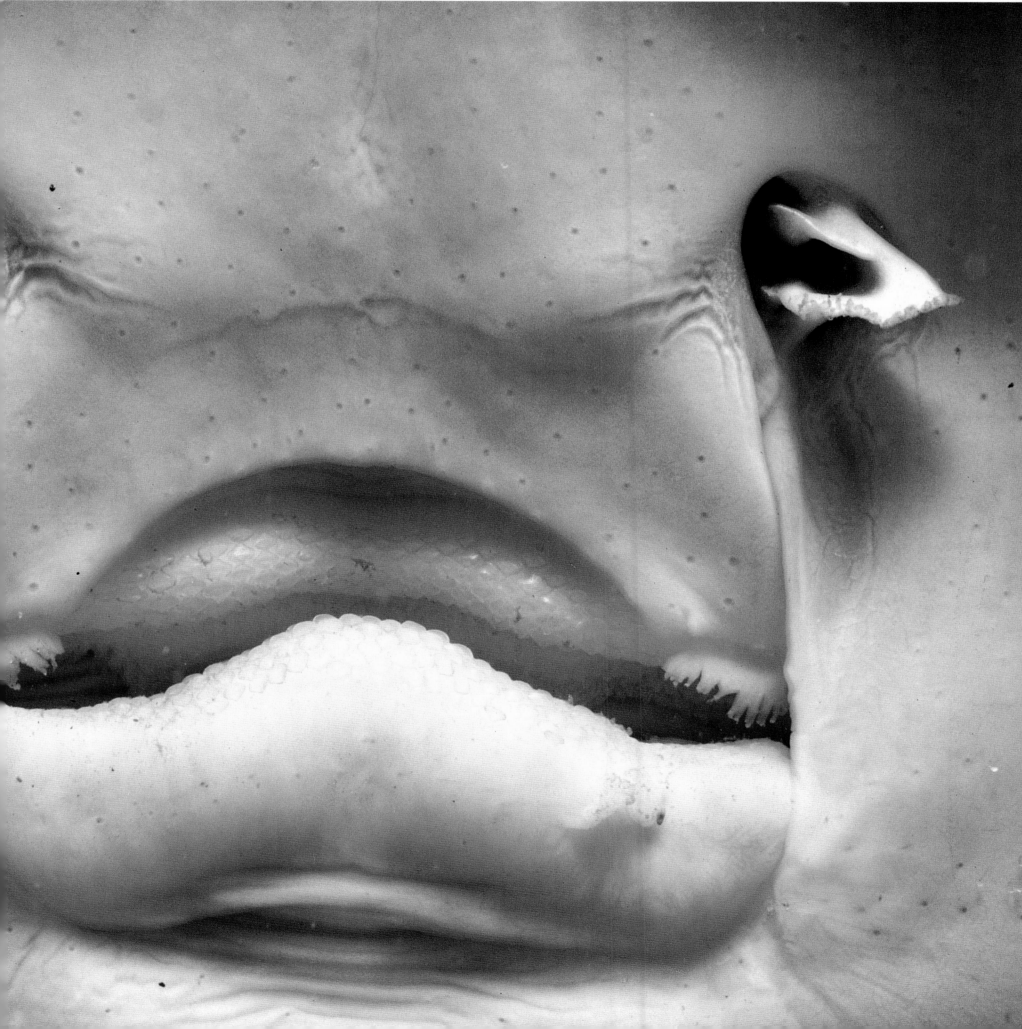

CATHEDRALS

Pages 88-89
Fan coral, Caribbean Sea.

Left
Gorgonian coral, Palau.

Corals are curious creatures. To the uninitiated, many resemble the sorts of improbable living rocks and animated flowers that science fiction writers envision as inhabitants of Venus. Yet the whole truth about corals is both more complex and more bizarre than any imaginary tale. Some—the hard corals—exist as a thin layer of living tissue stretched over intricately patterned limestone skeletons which form the bulk of the structures we call coral reefs. Others—the soft corals—do not assemble limestone skeletons like the reef builders; they create flexible supports that allow them to grow into an amazing variety of shapes.

Anyone who has observed corals for a few hours knows that they are animals. Though firmly rooted in place, corals do feed, and are composed of hundreds—or even thousands—of individual polyps, each of which bears a mouth and a set of tentacles. Corals' animal nature shows itself in other ways too; many species wage war on one another as they battle for space on the crowded reef. They do so, not in the manner of plants that merely scramble to overgrow each other, but like animals that attack, kill, and eat any neighbors that infringe on their turf!

On many reefs, hard corals are the most conspicuous of the group, because they grow best in the sunlit, wave-churned waters near the crest of the reef, where divers are most apt to see them. These corals' preference for brightly lit spots gives a clue to their nature, for while they are not plants, neither are they "simply" animals. For as the epitome of symbiotic associations in the sea, these organisms manage to function as both plant and animal at the same time. Precisely lined up within the tissues of the coral animals are thousands upon thousands of symbiotic guests; special forms of a single-celled alga that is also found swimming around in the open water of the reef. And though we, for the sake of convenience, may talk of coral animals and their plant partners as though they were separate beings, in actuality the two function together in ways that neither alone could manage.

By day, coral colonies do, in fact, function very much like plants. As they bask in the intense tropical sunshine, algae within coral tissues use that energy to assemble a variety of complex molecules of life from simpler component parts. The coral polyps, in the meantime, withdraw into their skeletons, and effectively disappear; all that remains of them for daytime visitors to see are the intricate patterns of ridges, grooves, and pores on the rocklike skeletons.

But divers after dark see a reef so transformed as to be almost unrecognizable. For at dusk, coral colonies metamorphose from passive sunworshippers to active predators. It is almost as though a field of wildflowers suddenly vanished, leaving in its place a menagerie of countless, hungry little animals. That's because coral polyps make their appearance during the twilight hours, expanding and spreading their tentacles in delicate crowns that look like living snowflakes scattered across the face of each colony. These fragile-looking polyps, however, carry a punch; their tentacles are covered with microscopic stinging cells which harpoon tiny animals that drift too close.

But the real wonder of this story is the network of intimate, invisible, chemical connections between corals and their symbiotic partners. Day and night, these plant and animal partners pass back and forth to one another a variety of compounds that make life easier for both of them. Animals provide the plants with natural fertilizer, making it possible for

Left
Gorgonian coral polyps, Caribbean Sea.

Below
Coral, Red Sea.

Right
Brittle sea star on Gorgonian coral, Caribbean Sea.

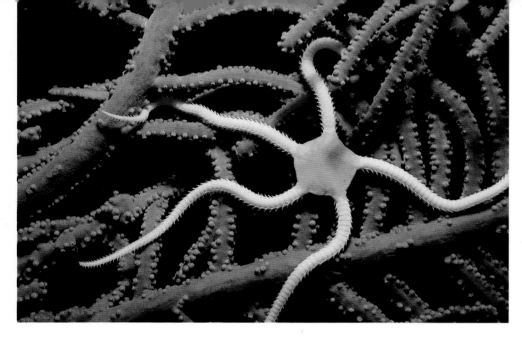

the algae to grow rapidly and efficiently. The plants, in return, provide their hosts with a varied menu of important chemical nutrients, supplementing their diet of plankton in vital ways.

The result of this convenient arrangement is a tightly knit cooperative unit, a sort of short-circuit ecosystem whose efficiency allows both members to grow far more luxuriantly than either could alone. The relationship benefits far more than just the corals themselves; without hard corals, the entire complex habitat of the coral reef, including all its marvelous fishes and invertebrates, would never be able to exist in the clear, almost nutrient-free waters of many tropical seas.

The only drawback to corals' partnership with algae is predictable; because the corals grow where their algal guests can trap sunlight, corals, like plants, cannot survive without light. Thus, hard corals are forever banished from the darkened depths of the sea and from the sunless caves, crevices, and overhangs within and beneath the very reefs they create.

But though the darker spaces are off-limits to hard corals, they are hardly devoid of life. For while hard corals are the conspicuous heart of the coral kingdom, they are but a fraction of a much larger empire that stretches from the ocean's surface to its darkest depths, and from the equator to the Arctic Circle. Most of that kingdom is ruled by animals commonly called soft corals, known to scientists as octocorals because their polyps bear eight tentacles.

Soft corals, which manage to survive without the aid of algal partners, build flexible skeletons instead of rigid ones. Some assemble supple, yet substantial, branches of material similar in structure to human hair; the dried, stiffened cores of these colonies are often made into jewelry. (In some parts of the world, in fact, the so-called "black corals" are nearly extinct because they have been overcollected by zealous jewelry producers.) Other soft corals make do with a loose network of tough, needlelike structures embedded in a body wall that may either be tough and leathery or have the consistency of day-old Jell-O.

In shape, soft corals vary wildly; some form huge, treelike masses, others produce miniature fans, curlicues, or even stems that resemble walking sticks. Because they don't depend on algal partners, soft corals grow easily in the dark. As a result, many species light up the deeper ledges and innermost recesses of the reef with colors that span the spectrum from violet to fluorescent green, bright orange, and deepest red.

As sedentary creatures in competitive seas, many soft coral species have evolved powerful chemical defenses to protect themselves. Some contain powerful antibiotics that apparently help prevent them from being covered with the bacteria that are so abundant in the sea. Others, like many land plants, contain bitter-tasting compounds that may help deter potential predators. Although relatively little is known about soft corals' arsenal of chemical defenses, researchers hope to co-opt some of these compounds to fight bacteria, viruses, and possibly cancer in humans.

But regardless of corals' potential economic appeal, their beauty and importance as bulwarks of tropical nature are indisputable. As the photographs on these pages show, they provide nearly endless variations of color, form, and texture, no matter where they are from or what time of day they are seen.

Pages 96 and 97
Alcyonarian corals, Red Sea.

Pages 98-99
Alcyonarian coral polyps, Red Sea.

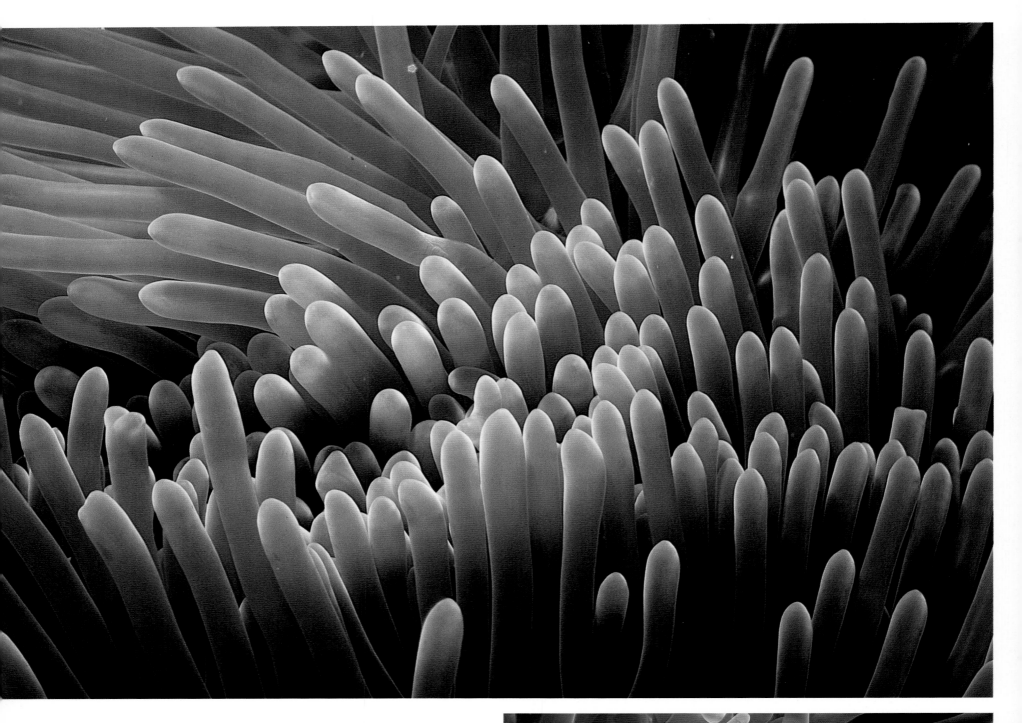

Above, right, and facing
Anemones, Palau.

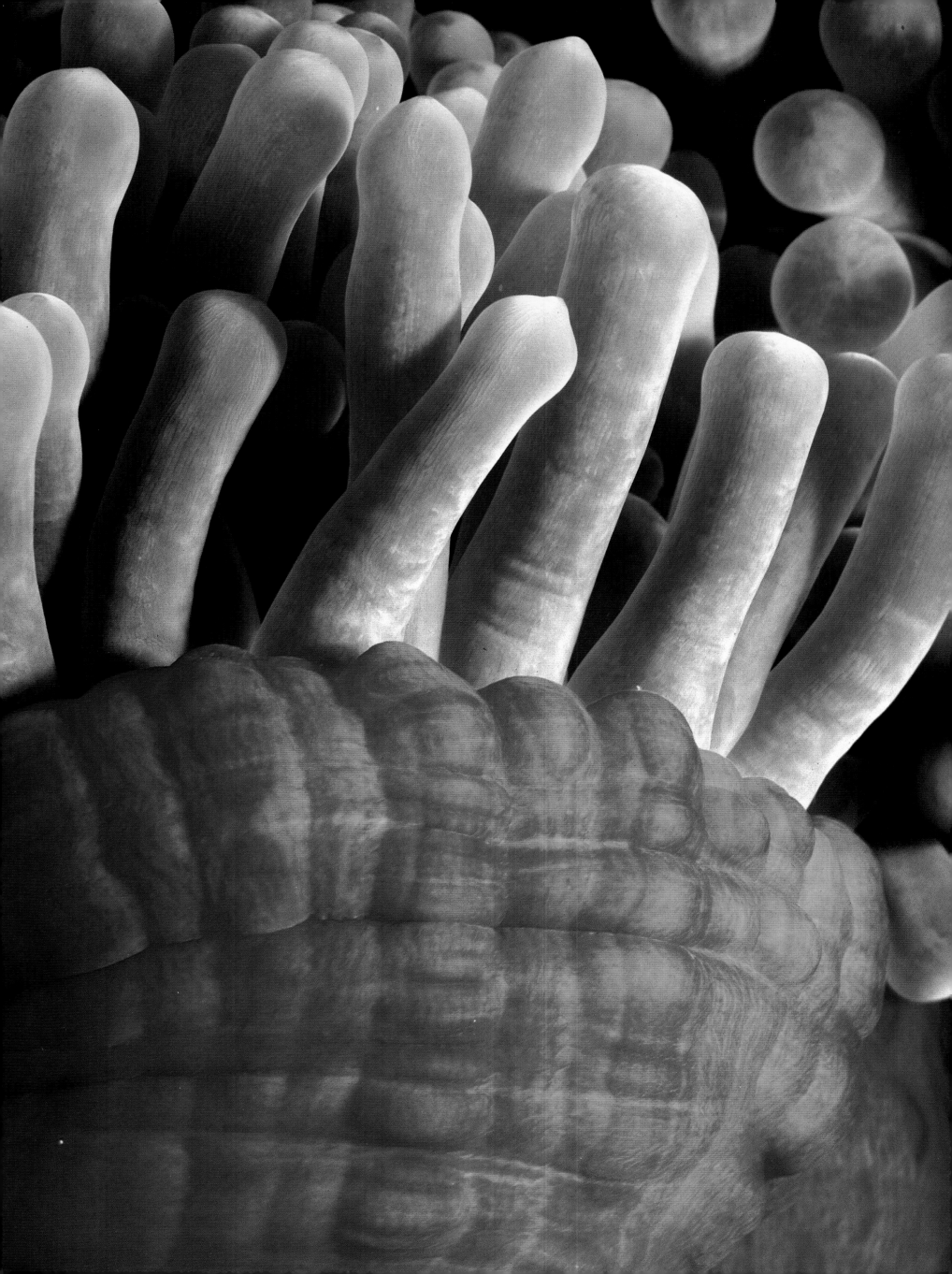

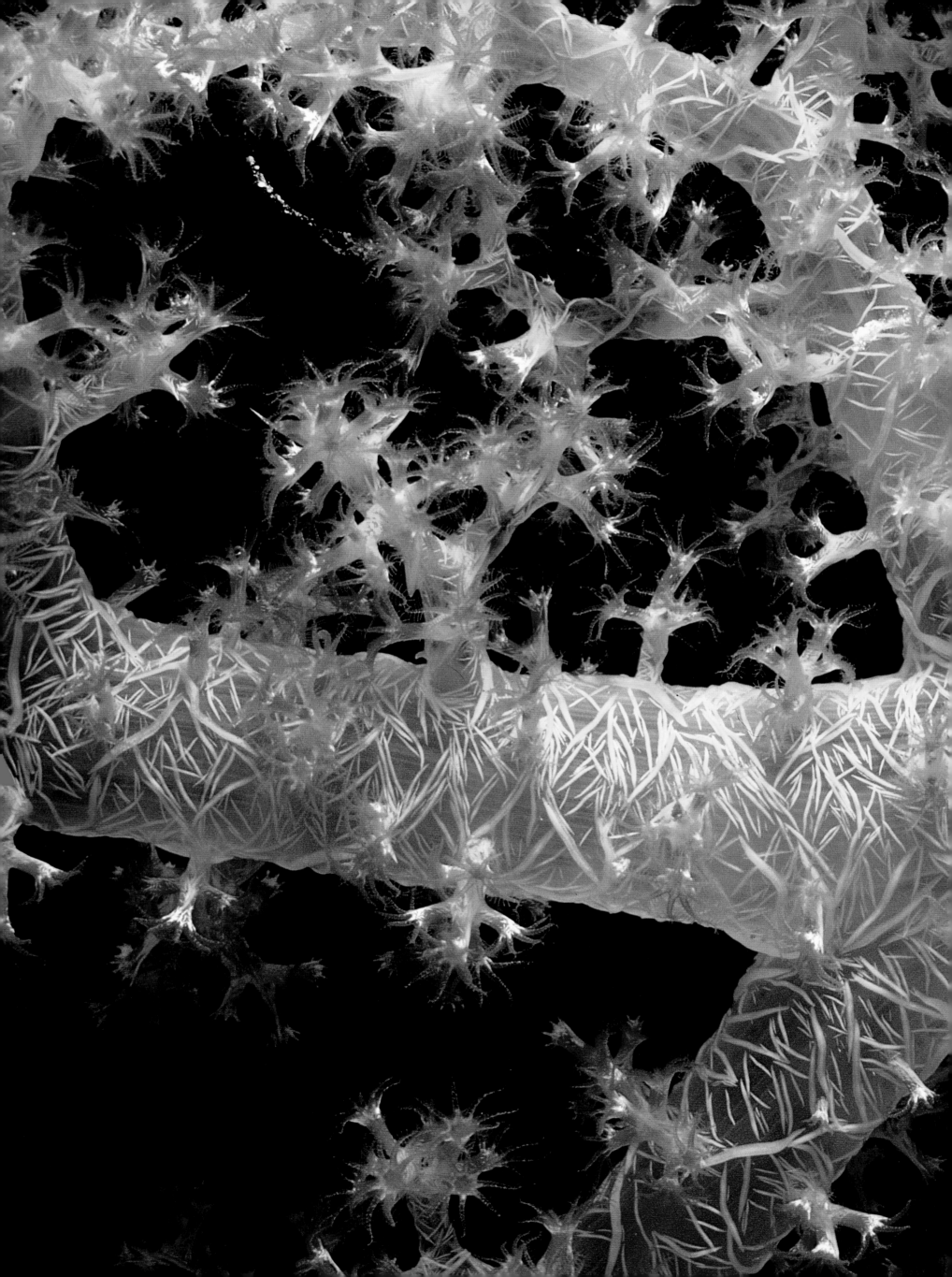

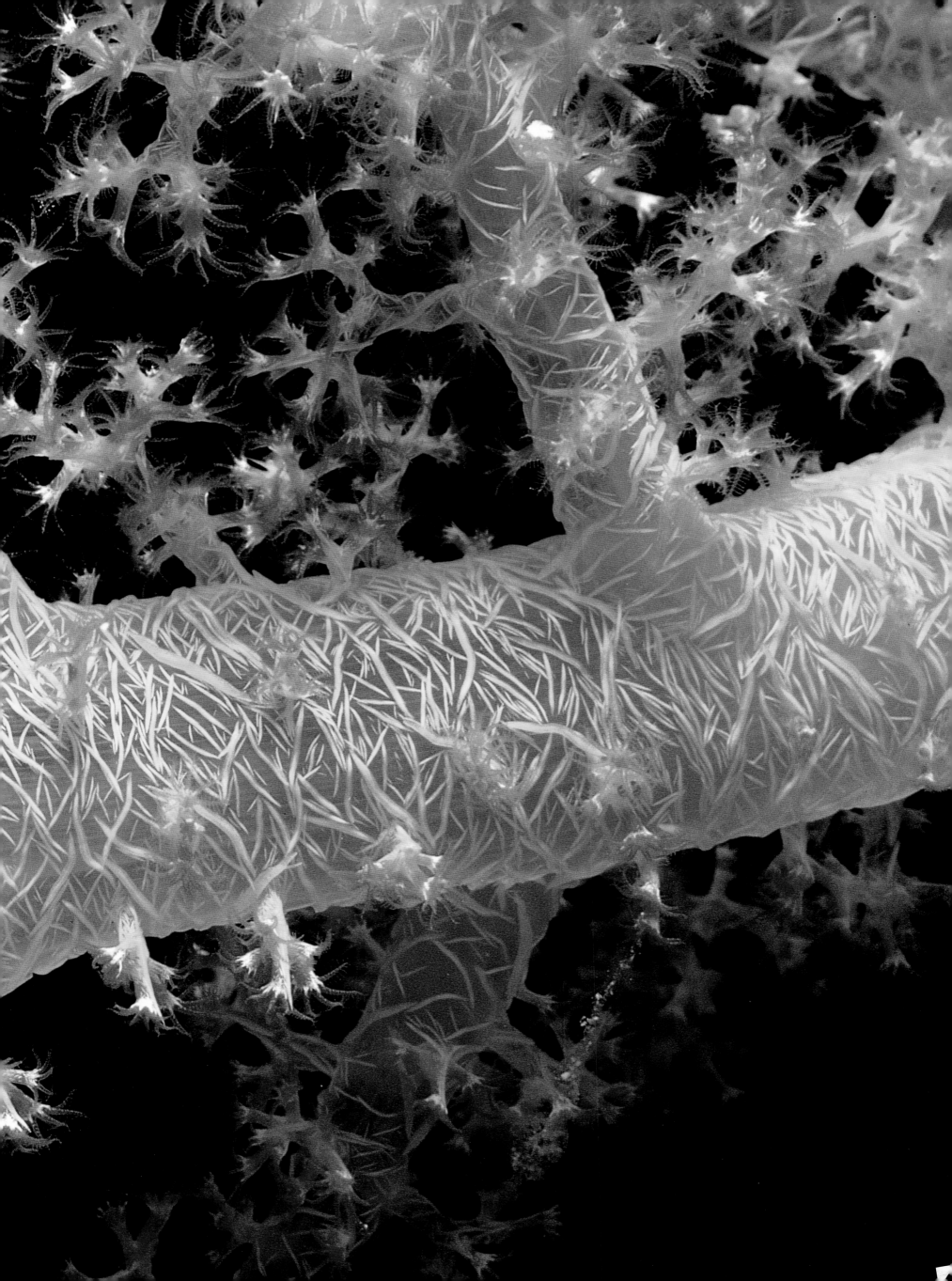

Above and facing
Gorgonian corals, Palau.

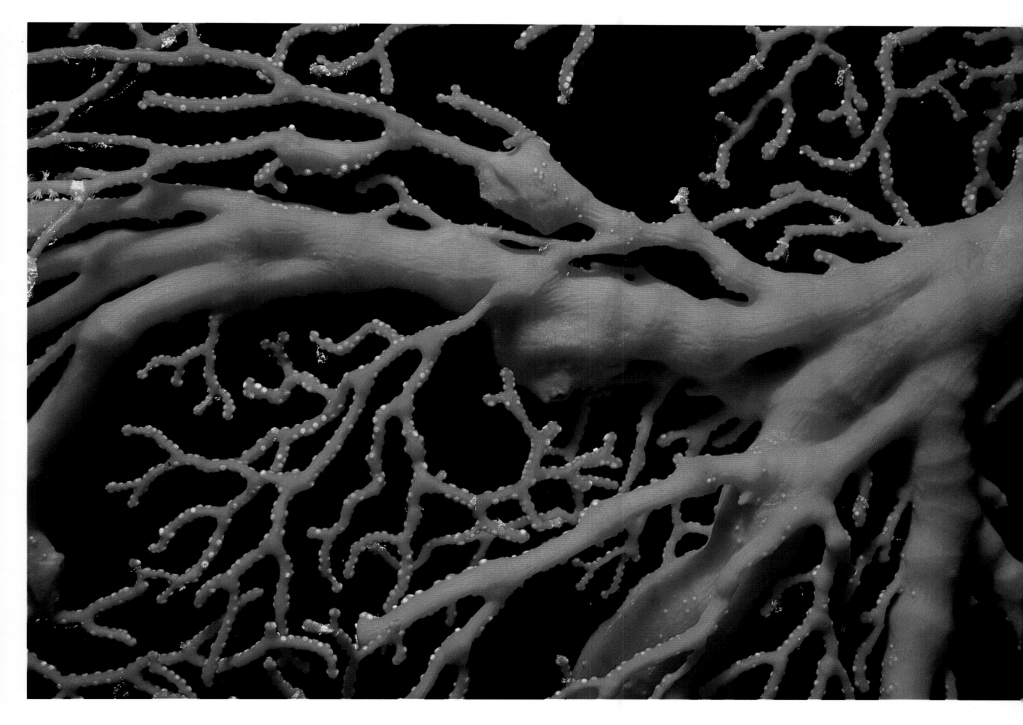

Pages 102-103
Gorgonian coral polyps, Red Sea.

Left
Brain coral, Caribbean Sea.

Below
Galaxea coral, Palau.

Pages 106-107
Symbiotic shrimp in the tentacles
of an anemone, Red Sea.

Pages 108-109
Tube coral polyps, Red Sea.

Left
Shrimp and anemone in symbiosis,
Red Sea.

Facing
Sponges and corals, Palau.

Pages 112-113
Bubble coral, Palau.

Pages 114-115
Stone coral, Red Sea.

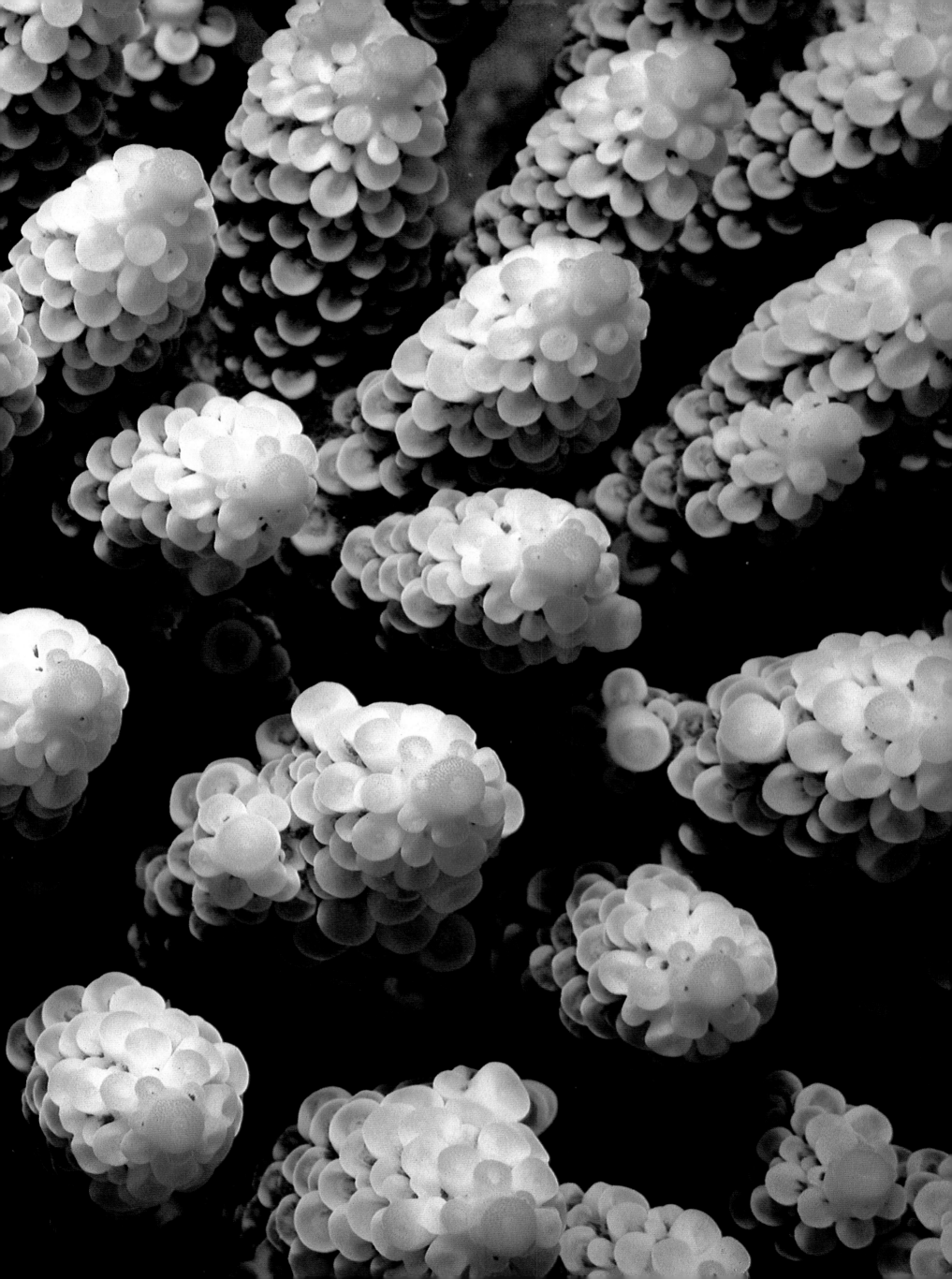

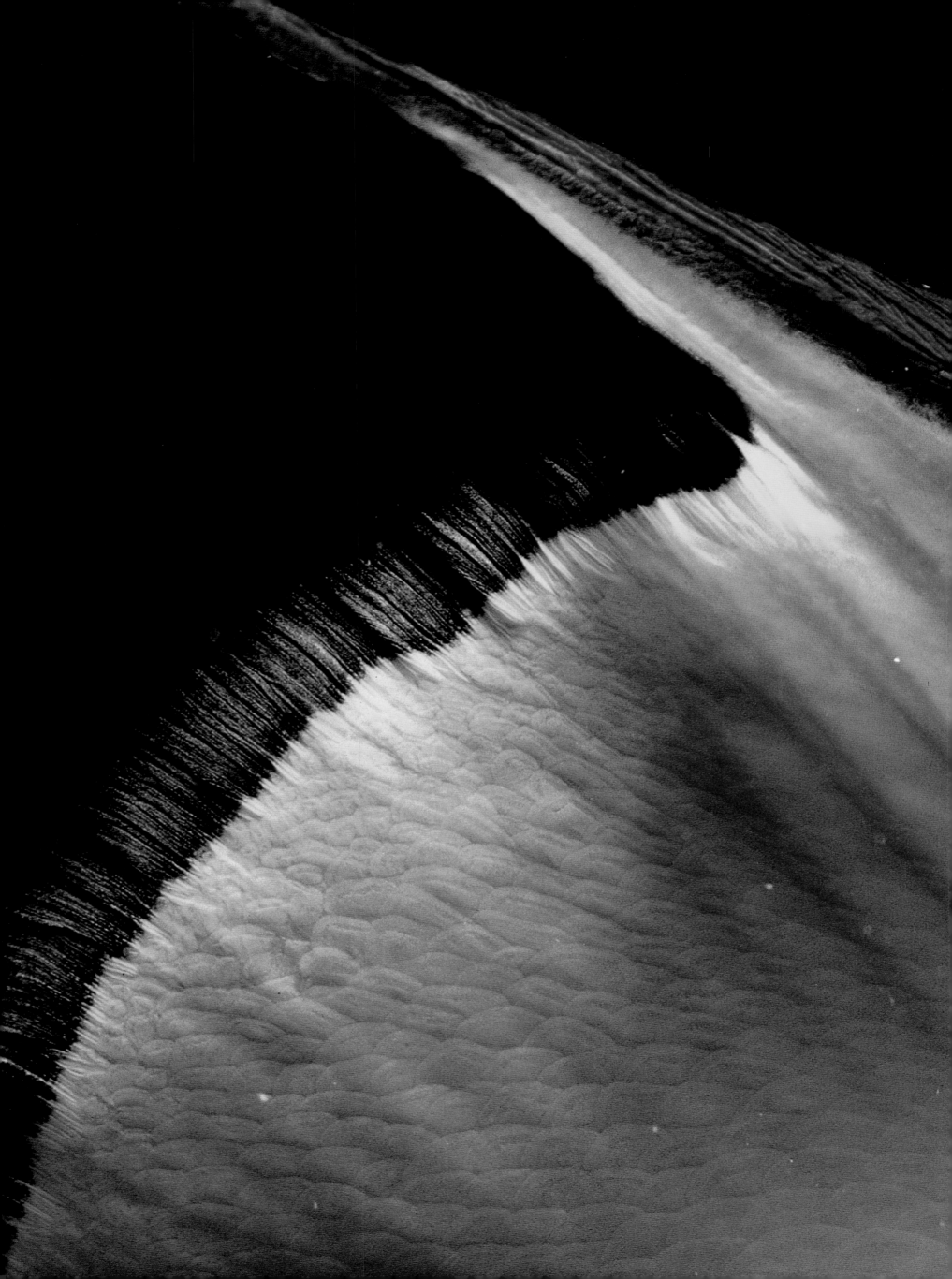

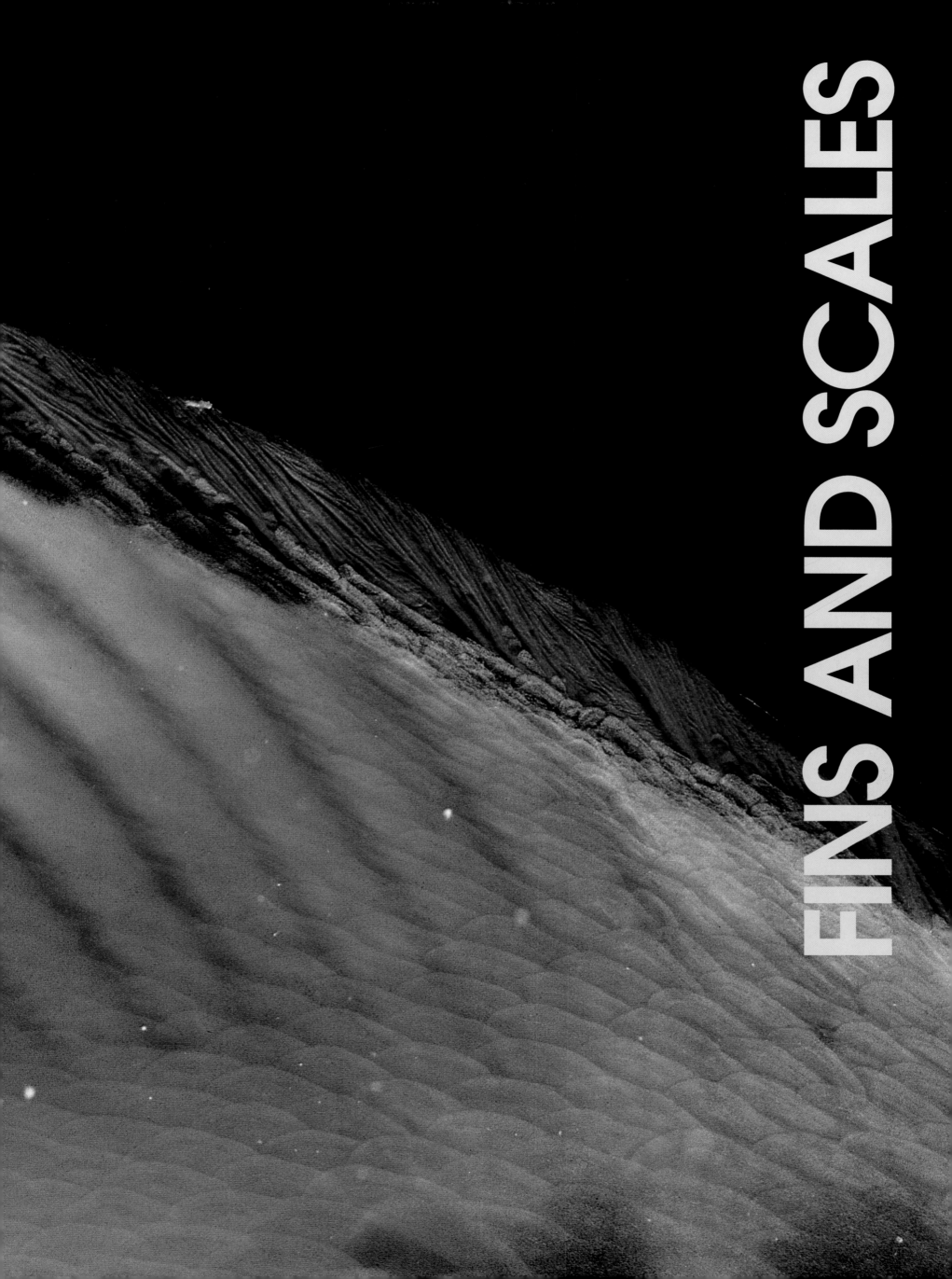

FINS AND SCALES

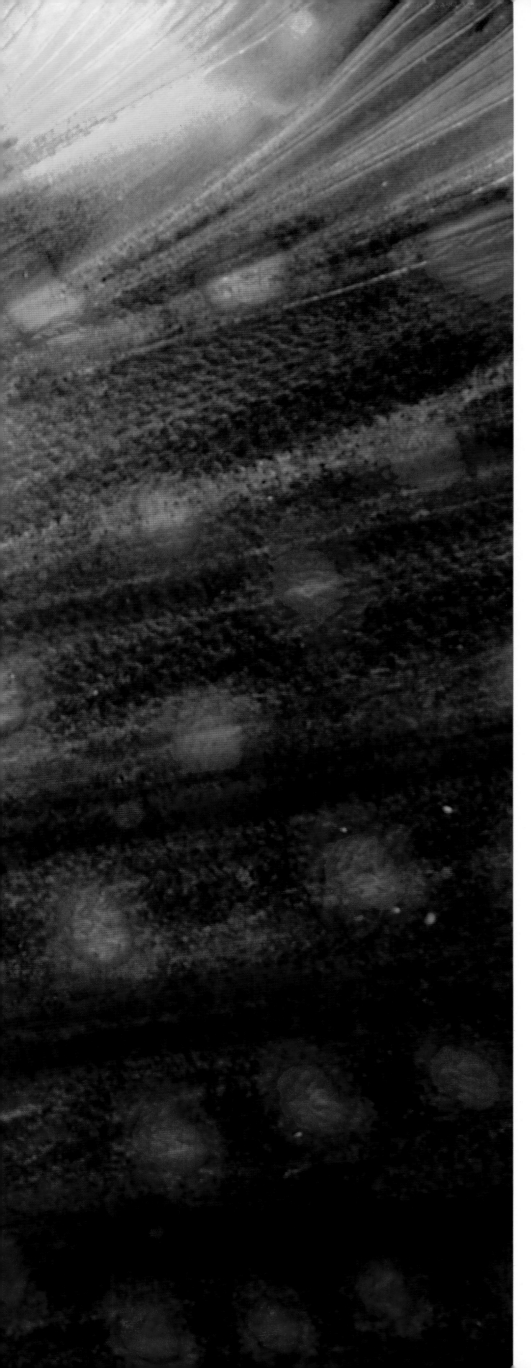

Pages 116-117
Butterflyfish, Red Sea.

Left
Jewel grouper, Red Sea.

Fish fins and scales play games with our eyes, attracting our attention here, diverting it there. Jeweled spots seem to vibrate against their backgrounds because they are ringed with a halo of contrasting brightness or hue. Fins are outlined, not just in white or black alone, but in combinations of white, black, and sometimes yellow and blue, as if to emphasize as much as possible the line where fish ends and empty water begins. Black eyes peer from within dark, concealing, raccoonlike masks that wrap around faces like shadows.

But the colors and designs on reef animals don't exist to amuse us; they evolved because each serves a useful purpose. Konrad Lorenz, Austrian scholar of animal behavior, described many of those patterns as *Plakatfarben* or "poster colors." He chose that name because he believed that fishes' bright patterns served the same purpose as advertising billboards—to broadcast messages boldly for all to see. So when we dive on reefs, the fishes we see swimming and dancing around each other are holding "conversations" in "words" of color, form, and behavior. To fishes, these languages are part of daily life; to us they are exotic as Sanskrit.

But what sorts of conversations are fishes carrying on with these advertisements? Study fishes' colors and behaviors for a while, and you'll find that they use their colors to communicate with one another in ways that are sometimes truthful and sometimes deceitful, but almost always ingenious.

Those species that "lie" about their presence are the ones that need to hide, either to ambush their next meal or to avoid being made into a meal by something else. These species must blend in as closely as possible with whatever they happen to be resting on. That's no easy job, for the shapes, colors, and textures of rocks, beds of sand and gravel, or a multi-colored carpet of algae and corals are both varied and complex. To match those patterns, animals adopt an equally multicolored and highly textured look. For that reason, if you see these "camouflage-colored" species in an aquarium, or in photographs taken outside their natural environment, their colors and textures look beautiful, and anything but inconspicuous. But in their natural haunts, these hunters-by-stealth literally vanish against equally boldly patterned backgrounds, and look amazingly like pieces of the reef itself. Even experienced humans have trouble picking them out.

The fishes that catch and hold our eye as we swim over the reef, on the other hand, are those that use colors and patterns, not to conceal their identities, but to advertise, to convey as accurately as possible a variety of specific messages. Given the ecological diversity of reef animals, it isn't surprising that there is no single common function for color patterns among coral fishes.

Yet there does seem to be at least one common thread that ties together many specific reasons for the importance of communication on the reef: the need to be noticed and understood in a crowd. For the reef is a habitat that many researchers believe carries nearly as many animals as it can physically hold. Under those conditions, fishes' ability to stake out and defend their territory may be vital to their survival. Similarly, where hundreds of species coexist, accurate communication between males and females of a species is essential to their success at producing the next generation.

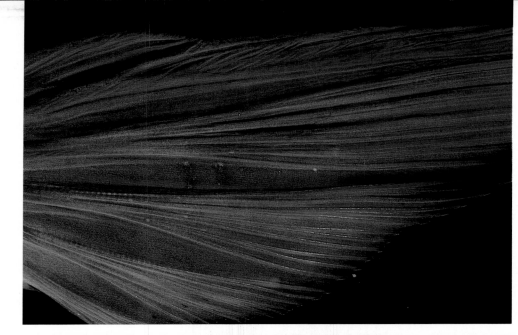

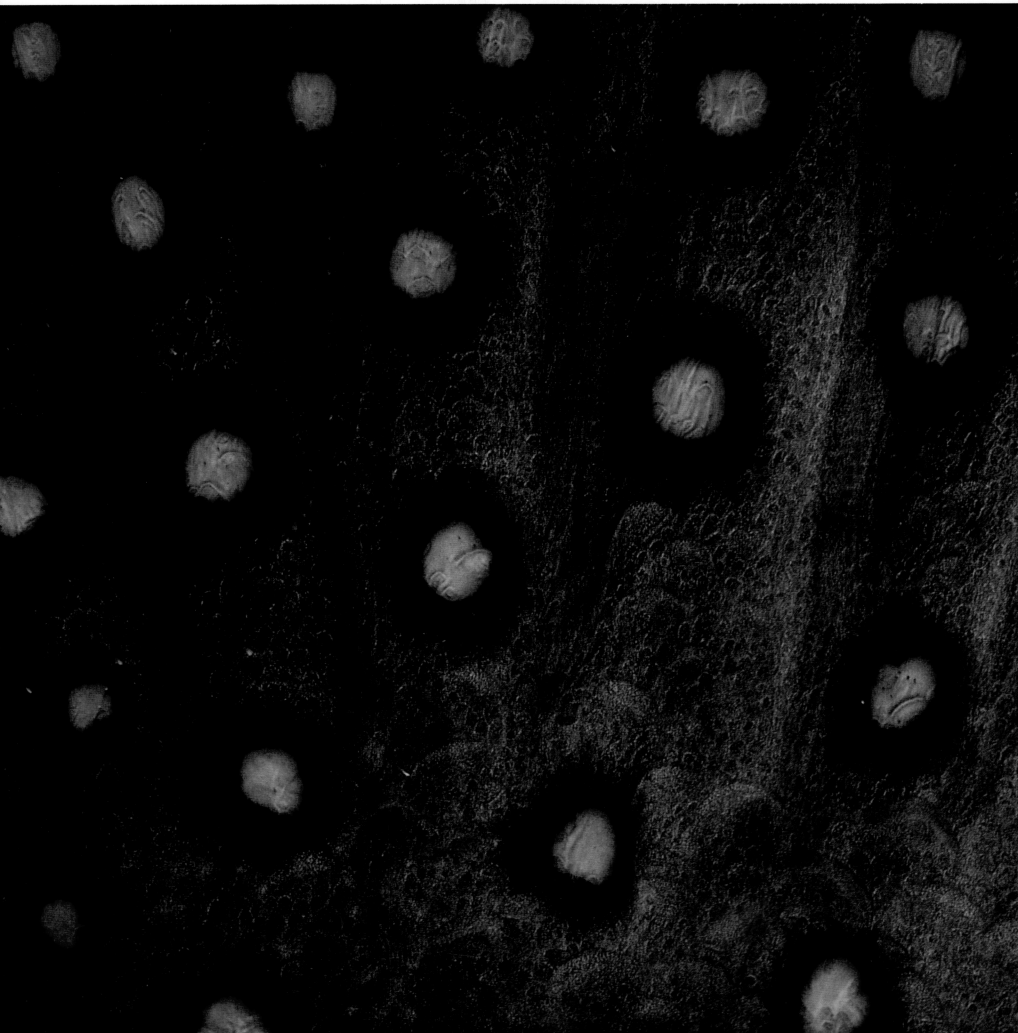

Left
Pectoral fin of a parrotfish, Red Sea.

Below
Leopard grouper, Red Sea.

Right
Emperor angelfish, Red Sea.

Pages 122-123
Yellowtail triggerfish, Red Sea.

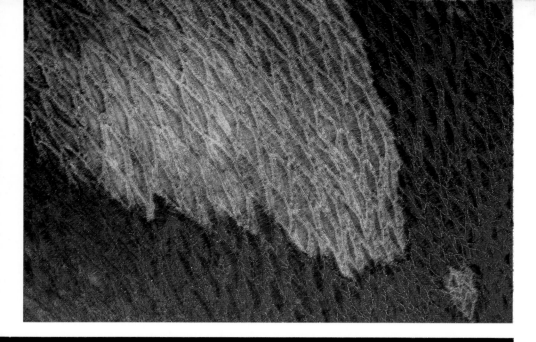

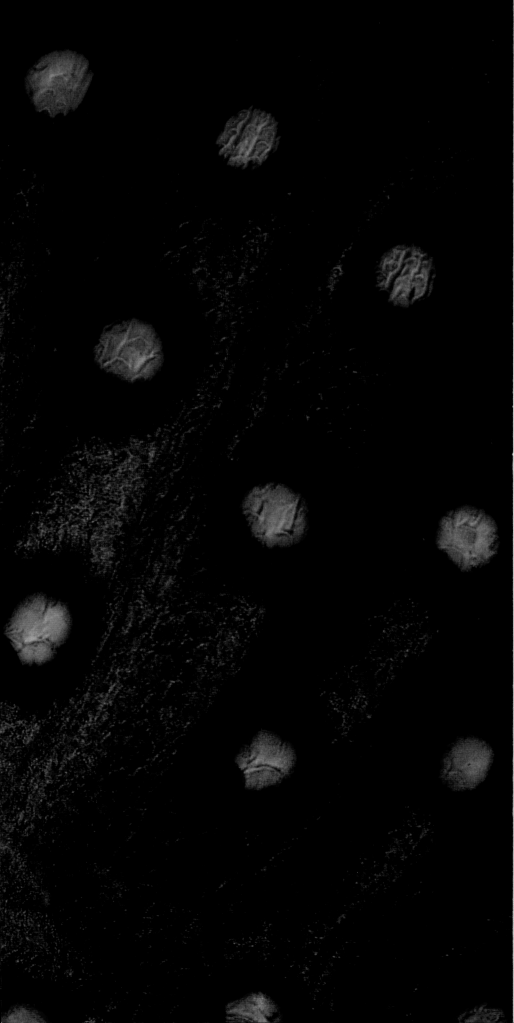

In several species of damselfish, for example, certain colors may signal success or defeat in battle or announce sovereignty over a patch of turf. Among groupers, striking color changes may be used to register alarm or signal submission after a fight. Among wrasses and parrotfishes, distinct color patterns on males help identify them to females at mating time. And the bright markings of butterflyfishes seem to help members of mated pairs keep in contact while travelling around to feed. In all of these cases, the urgency of the message requires that it be highly visible in the visually complex coral realm.

To achieve these goals, fishes mix colors and patterns with astonishing effectiveness. Intriguingly, the forces of evolution—that combination of chance and natural selection—have molded fish colors and patterns into many of the same combinations used by human artists to catch and hold the mind's eye or to create illusions of size, depth, or movement. Bold stripes, bars, dots, and chevrons of black on white appear everywhere on the reef, attracting and holding the eye as effectively as the most effective human advertisement. Brilliant yellow, the most visible of all colors in tropical seas, is among the most commonly used colors on the reef. That yellow is often paired with saturated blue—a near-perfect complement—to create maximum contrasts as extreme as those generated by the most garish neon signs.

But unlike human paintings, many fish compositions are dynamic and can change dramatically with the animals' moods. How can fishes manage such tricks? Fish body colors are created by pigments within specialized skin cells. Many of those pigments come in the form of tiny granules that meander within large, branching cells. In less than a minute, a fish can order those granules to dance around, causing them either to clump together or to disperse. When the granules cluster, much light passes around them, rebounds from the white background, and reflects back to our eyes. Under these conditions, the skin appears light. On the other hand, if a dark pigment, melanin, is spread throughout the cells like a shroud, it traps much of the light that hits it, darkening the skin.

Still, one general, nagging question remains. Elsewhere in the aquatic world (and on land, too, for that matter) the threat of predation seems to have forced most animals into hiding, rather than advertising their whereabouts. There are, after all, neither fluorescent-yellow herring nor red-and-purple antelopes. So how can reef fishes use vivid colors brazenly to proclaim their existence for everyone—friend and foe—to see?

The answer preferred by many researchers brings our thoughts around full circle, back to the reef itself, where we began our journey. Remember that the complex honeycomb of old coral skeletons affords an almost infinite number of hiding places for fishes swimming near it. Unlike a fish in open water, a reef fish that spies a predator has a good chance of finding shelter within a very few strokes of its tail. Remember, too, that most fish-eating predators on the reef, including groupers and sharks, hunt most actively at dawn and dusk, when bright colors are less visible anyway. It may thus be that reef fishes' colors are less important than reasonable speed and agility in keeping them out of the jaws of death. Given the importance of keeping their messages straight, the benefits of bright colors won out, to the benefit of reef animals and the continued fascination of those of us who visit their world.

Above
Pectoral fin of a jewel grouper, Red Sea.

Above
Pectoral fin of a wrasse, Red Sea.

Pages 126-127
Dorsal fin of a surgeonfish, Red Sea.

Page 128
Emperor angelfish, Red Sea.

Page 129
Orange-striped triggerfish, Red Sea.

Pages 130-131
Butterflyfish, Red Sea.

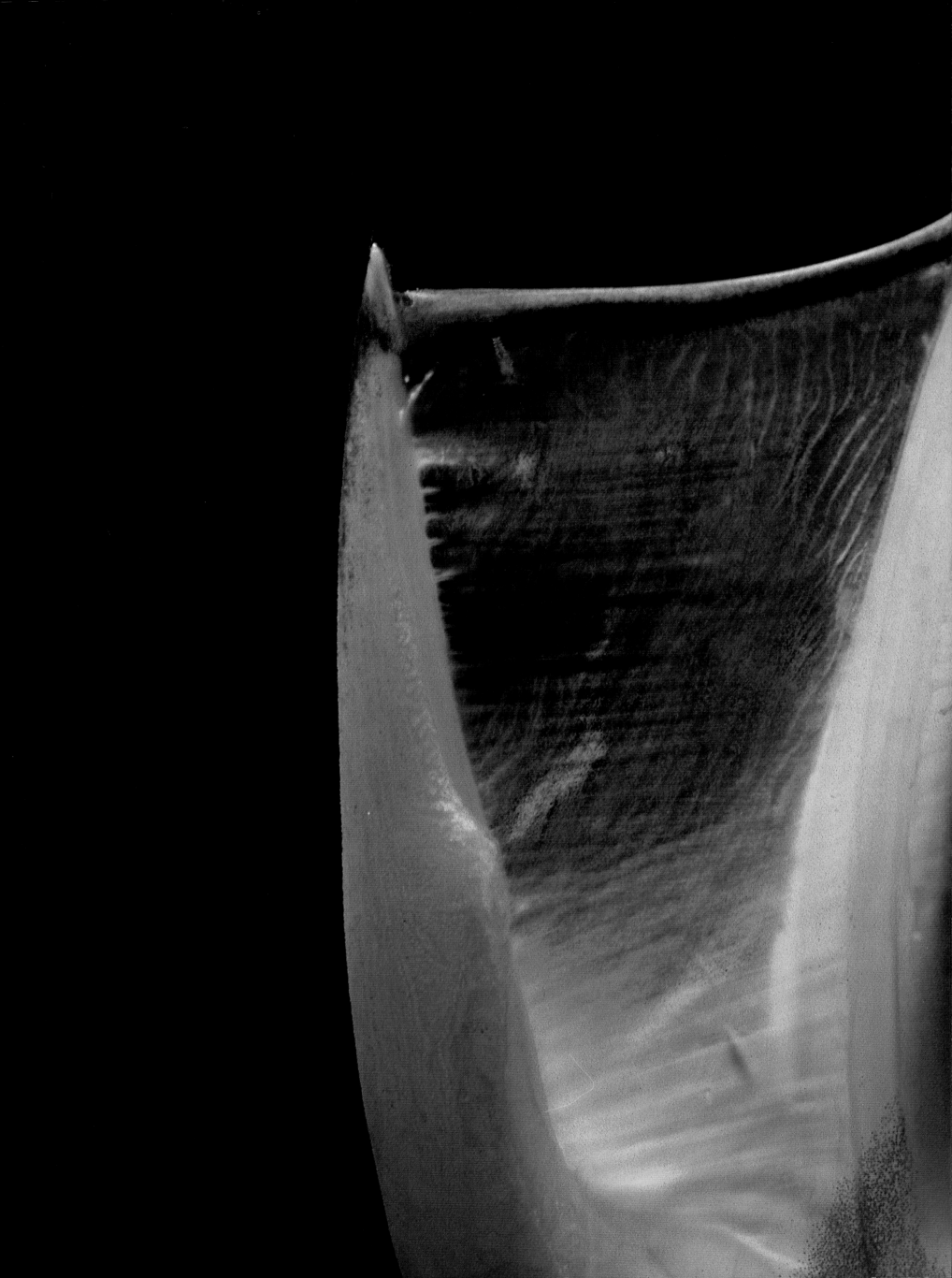

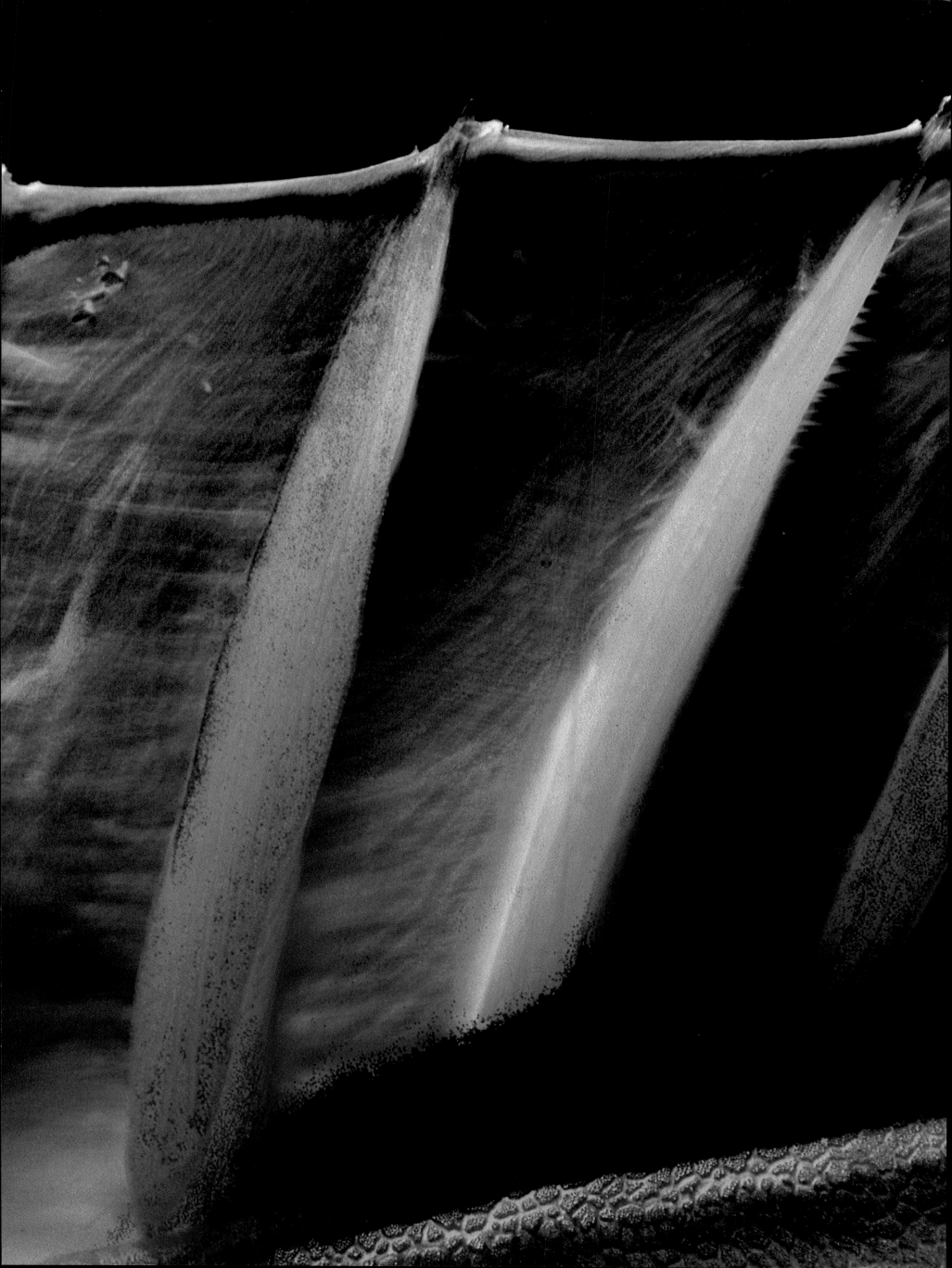

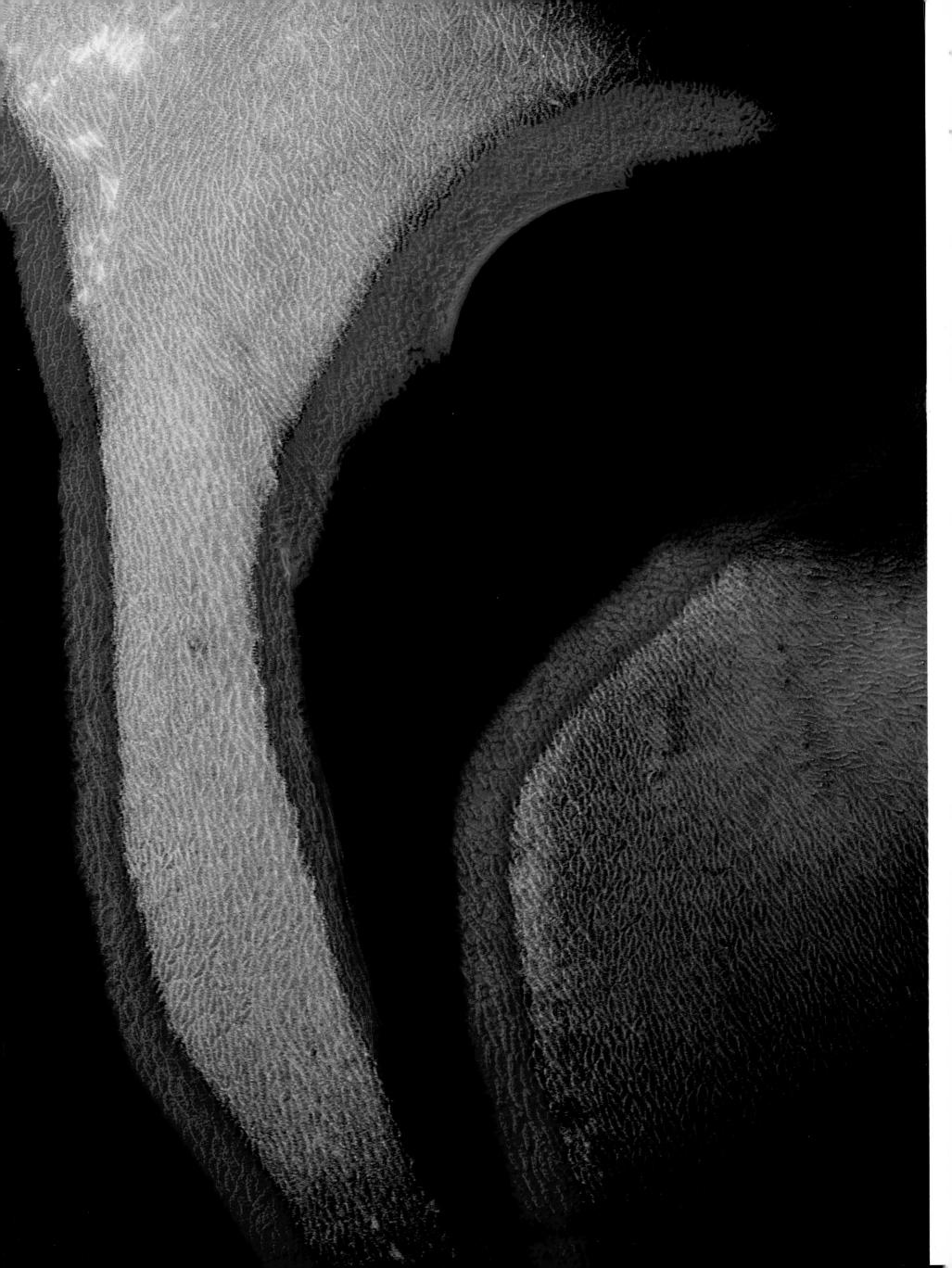

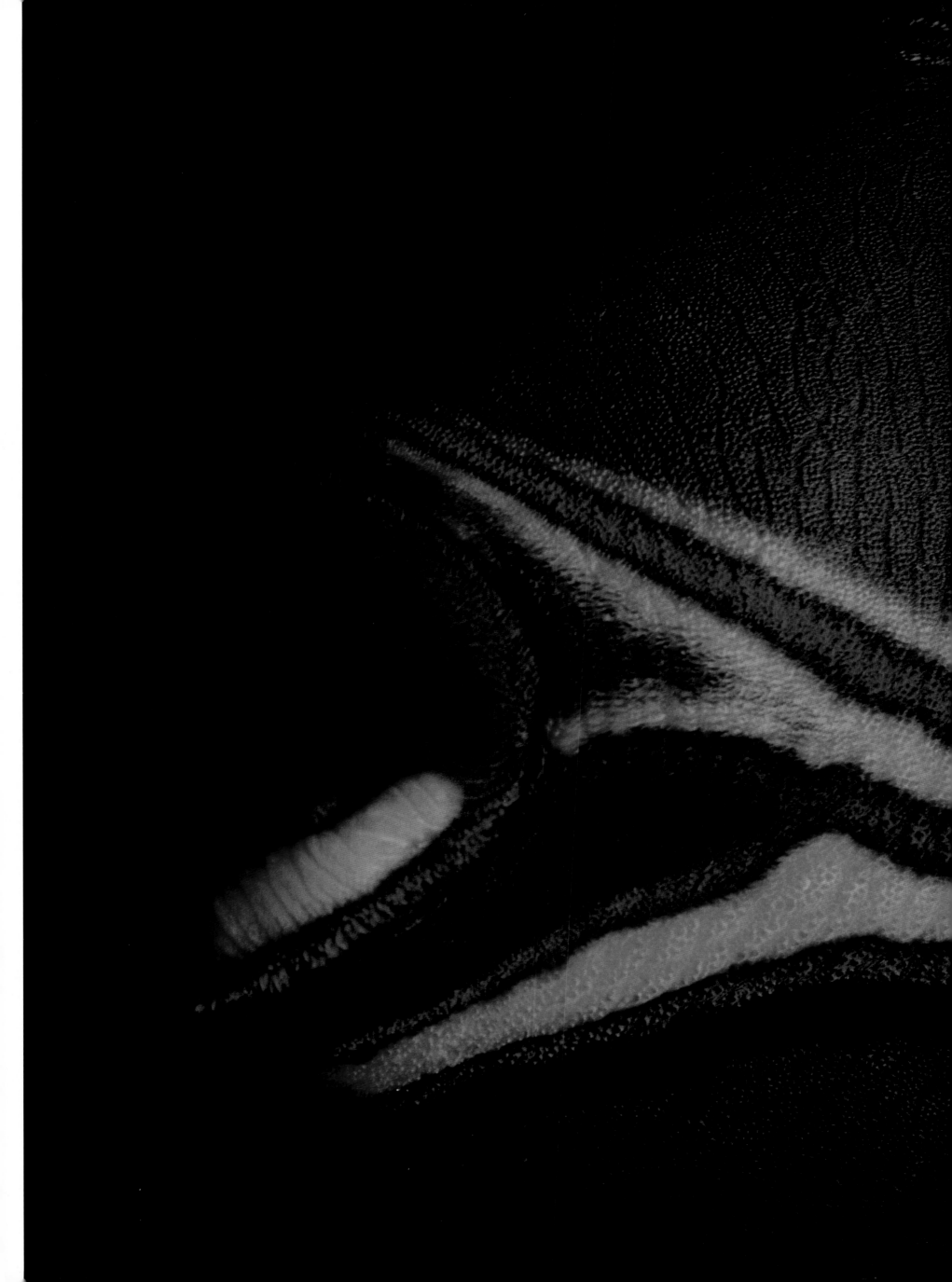

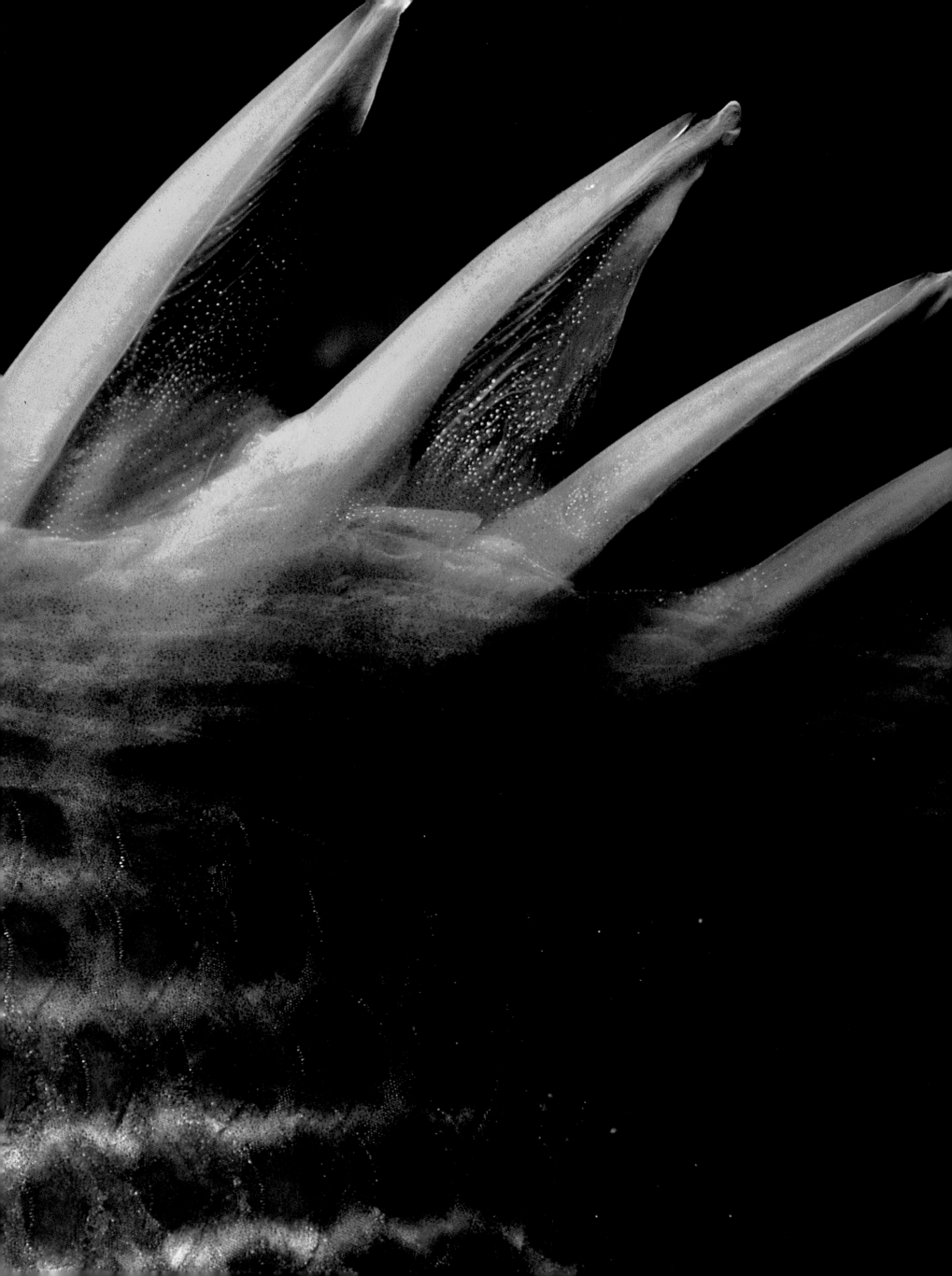

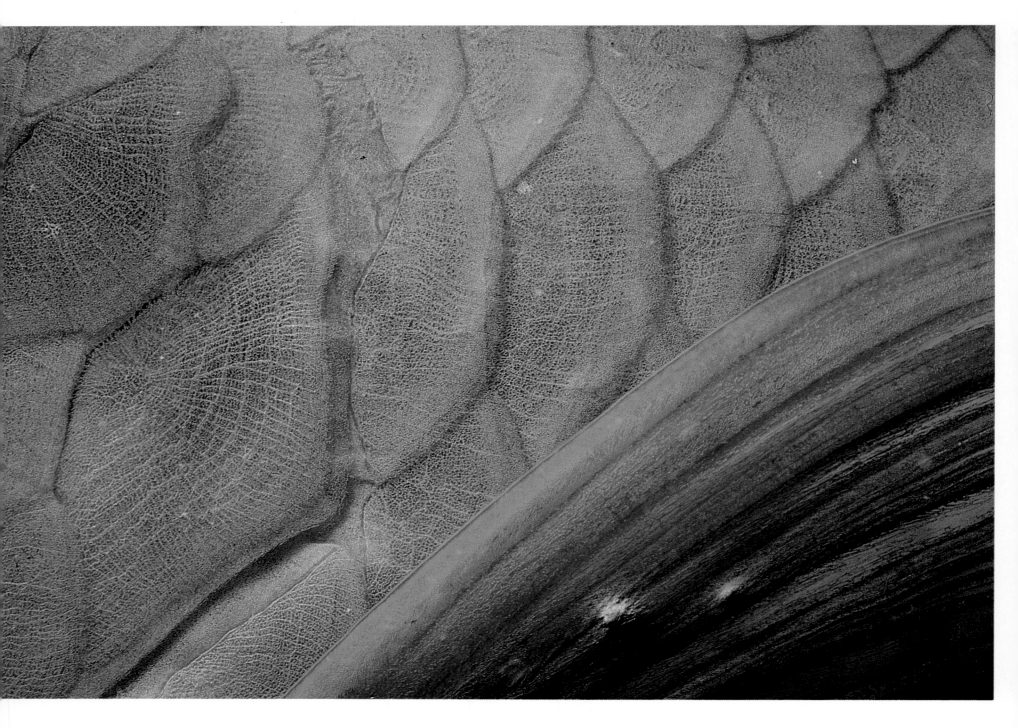

Above
Rusty parrotfish, Red Sea.

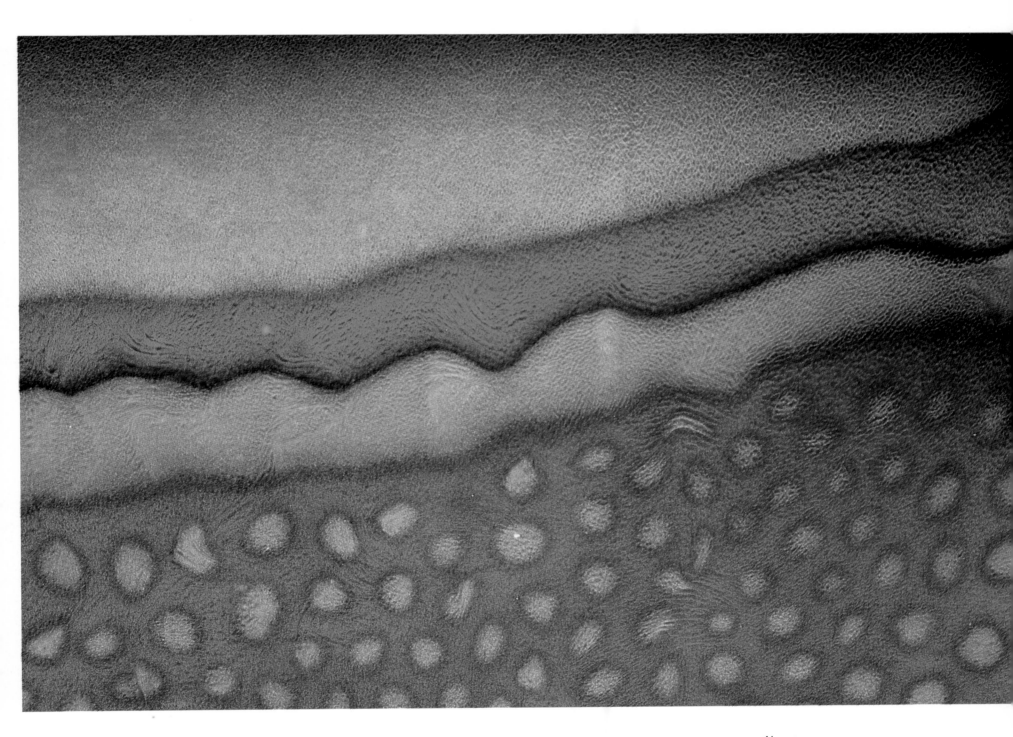

Above
Parrotfish, Red Sea.

Pages 134-135
Crown-of-thorns sea star, Caribbean Sea.

Pages 136-137
Fire lionfish, Red Sea.

Pages 138-139
Scorpionfish, Red Sea.

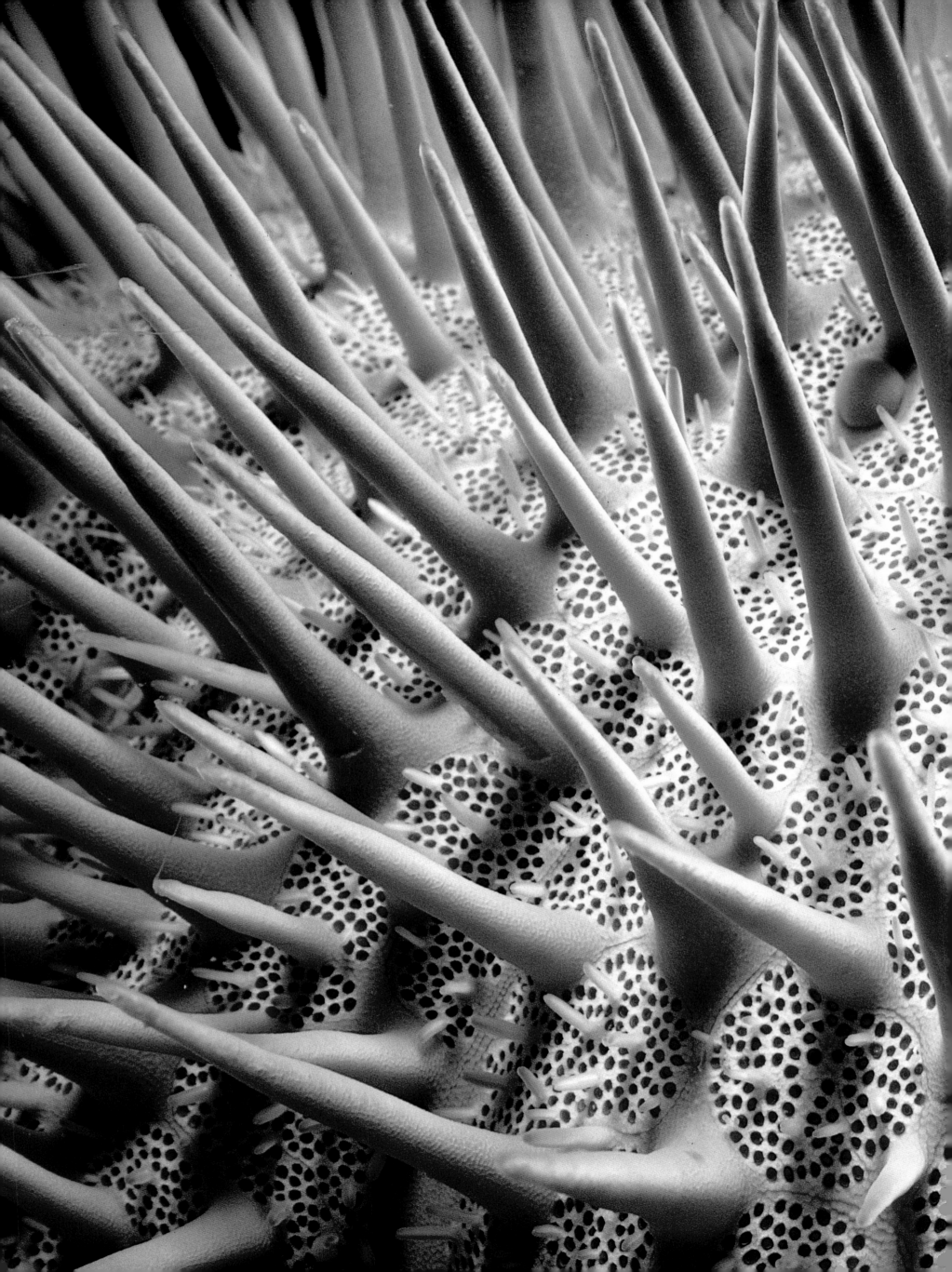

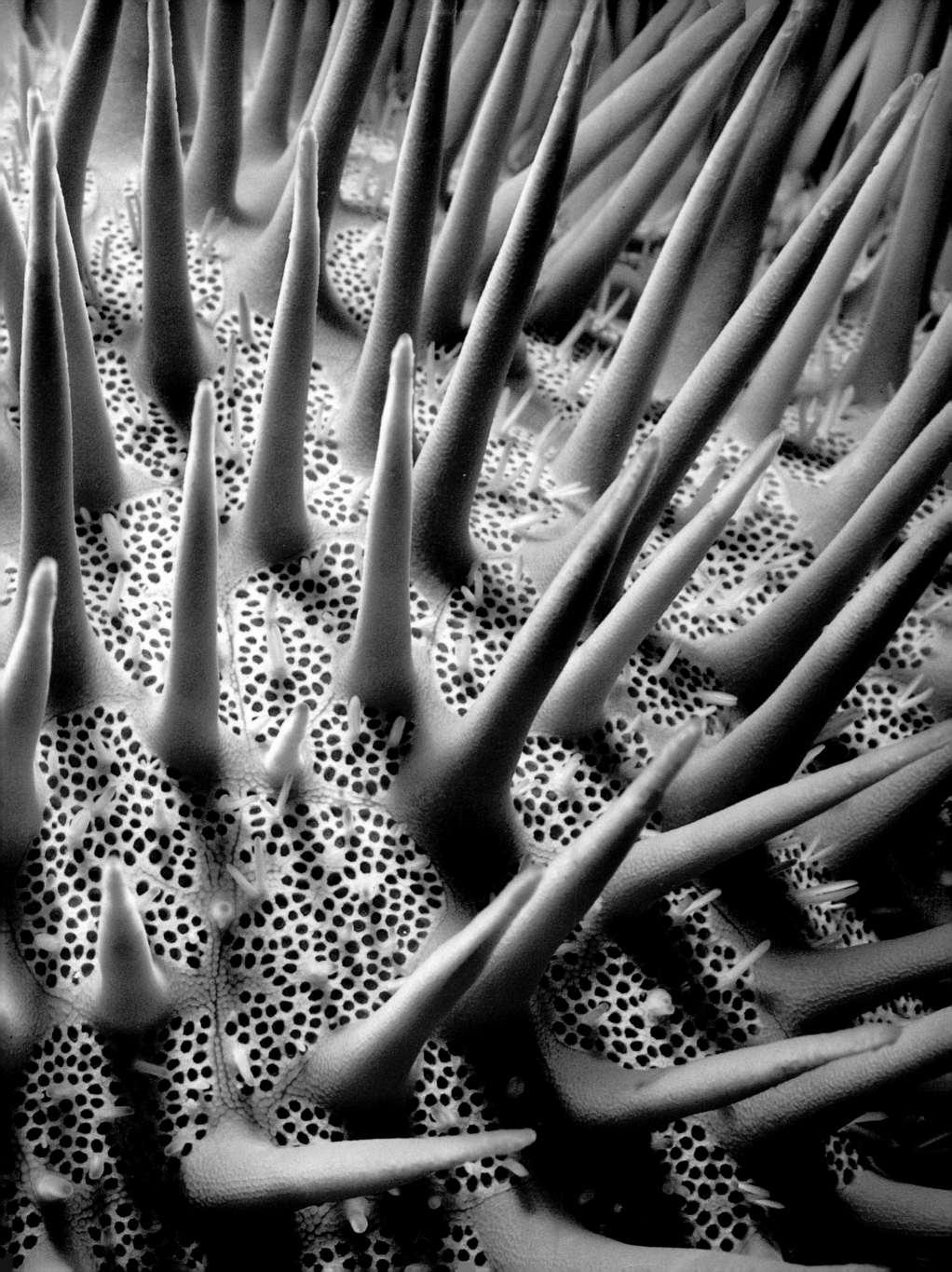

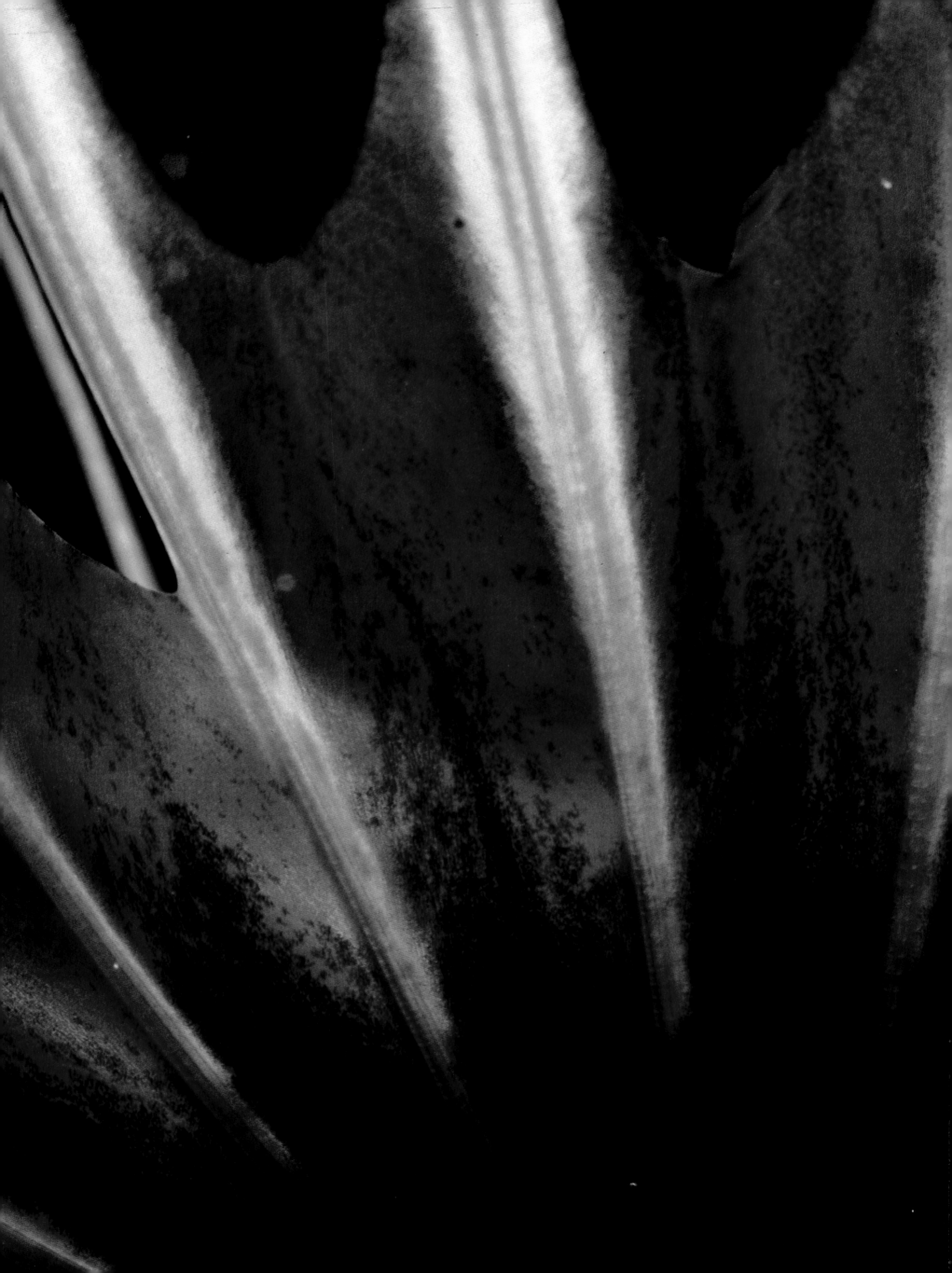

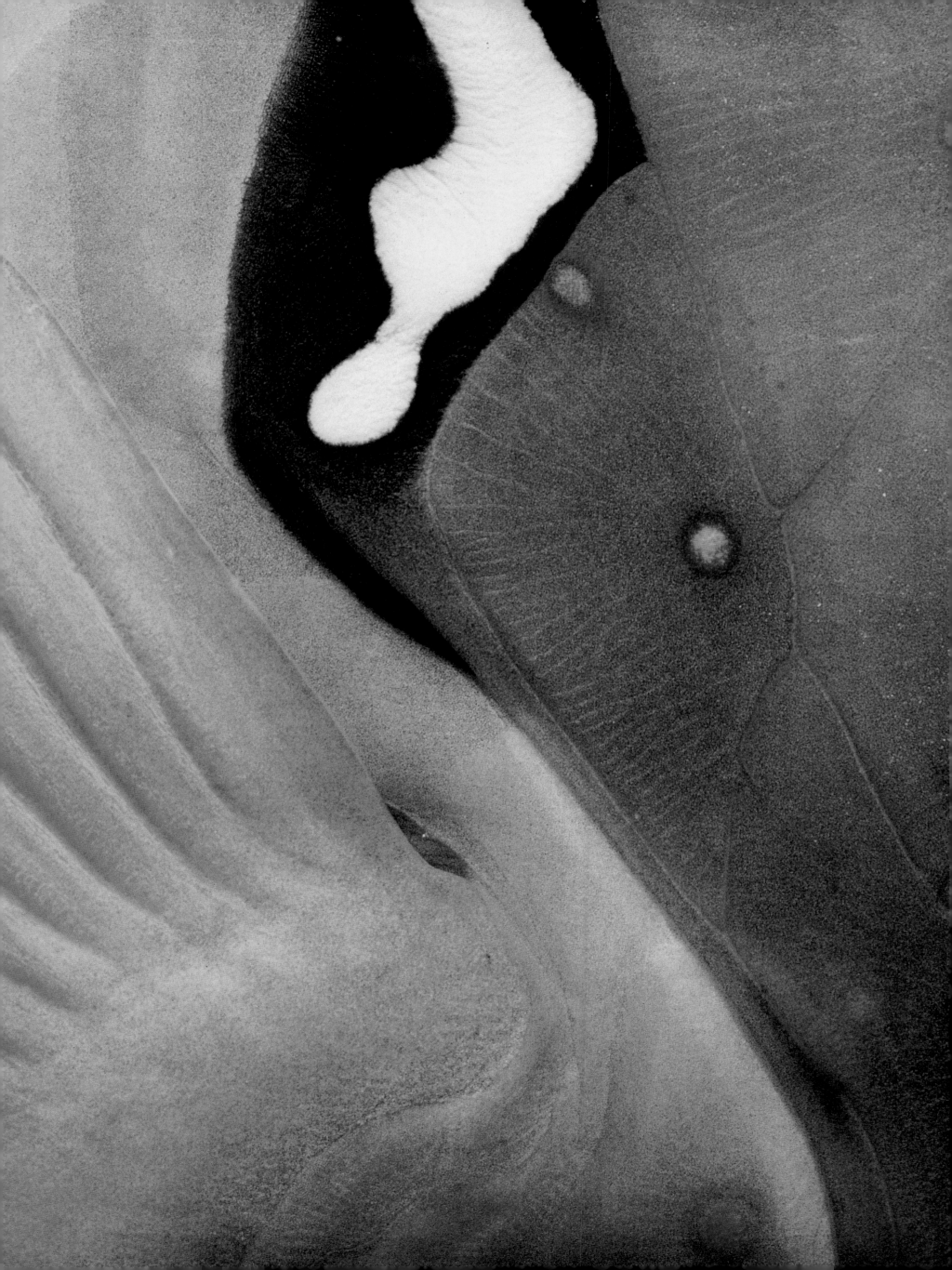

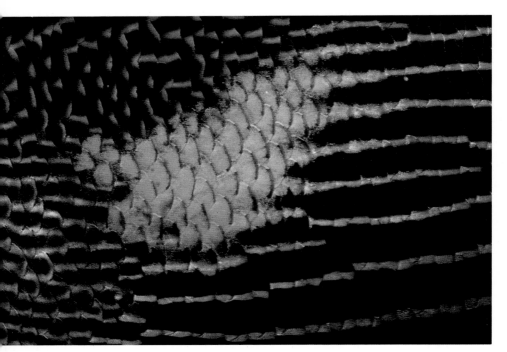

ACKNOWLEDGMENTS

Underwater photography is not created in solitude, and this book would not have been possible without the friendship and help of the following people:

Jane and Bob Altman, Peter Arnold, Mark Ausenda, Fred Bavendam, Ken Beck, Pascal Bessaoud, Boris-Master Color, Bob Boyle, Bob Braun, Christiane Breustedt, Giovanna Calvenzi, Howard Chapnick, Natasha Chassagne, Caroline Despard, Lydie Driviere, Embarak of Sinai, Robert Fiess, Richard Fraiman, Steve Freligh, David Friedman, Gianluigi Gonano, Amos Goren, Peter Gorman, Itamar Greenberg, Muhamed Hagrass, Willy Halpert, Yair Harel, Roy Hauser, Uta Henschel, Paul Humann, Bob Johnson, Venita Kaleps, Jon Kenfield, Moshe Kotler, Maria Lane, Joe Levine, Barbara London, Brian Lyons, Nadia Mancy, Ernst Meier, Urs Mockli, Koji Nakamura, John Nuhn, Lello Piazza, Sylvie Rebbot, Elke Ritterfeldt, Robert Rosenberg, Howard Rosenstein, Galit and Dana Rotman, Rene and Nate Rotman, Mark Schoene, Laurie Seluk, Mel Scott, Julio Stopnicki, Francis Toribiong, Manuel Velasco, Anne Vinnecombe, and Neal Watson.

Page 140
Yellowtail triggerfish, Red Sea.

Page 141
Surgeonfish, Red Sea.

Pages 142-143
Broomtail wrasse, Red Sea.

Above
Surgeonfish, Red Sea.